Environmental Issues

WATER POLLUTION

Environmental Issues

AIR QUALITY
CLIMATE CHANGE
CONSERVATION
ENVIRONMENTAL POLICY
WATER POLLUTION
WILDLIFE PROTECTION

Environmental Issues

WATER POLLUTION

Yael Calhoun
Series Editor

Foreword by David Seideman,
Editor-in-Chief, *Audubon* Magazine

CHELSEA HOUSE
PUBLISHERS
A Haights Cross Communications Company ®
Philadelphia

CHELSEA HOUSE PUBLISHERS

VP, NEW PRODUCT DEVELOPMENT   Sally Cheney
DIRECTOR OF PRODUCTION   Kim Shinners
CREATIVE MANAGER   Takeshi Takahashi
MANUFACTURING MANAGER   Diann Grasse

Staff for WATER POLLUTION
EXECUTIVE EDITOR   Tara Koellhoffer
EDITORIAL ASSISTANT   Kuorkor Dzani
PRODUCTION EDITOR   Noelle Nardone
PHOTO EDITOR   Sarah Bloom
SERIES AND COVER DESIGNER   Keith Trego
LAYOUT   21st Century Publishing and Communications, Inc.

A Haights Cross Communications ✦ Company ®

First Printing

9 8 7 6 5 4 3 2 1

Library of Congress Cataloging-in-Publication Data

Water pollution/[edited by Yael Calhoun]; foreword by David Seideman.
    p. cm.—(Environmental issues)
      Includes bibliographical references and index.
    ISBN 0-7910-8202-4
    1. Water—Pollution. I. Calhoun, Yael. II. Series.
TD420.W337 2005
363.739'4—dc22

                                                           2004028992

All links and web addresses were checked and verified to be correct at the time of
publication. Because of the dynamic nature of the web, some addresses and links
may have changed since publication and may no longer be valid.

CONTENTS OVERVIEW

Detailed Table of Contents

Foreword

by David Seideman, Editor-in-Chief, *Audubon* Magazine

For anyone contemplating the Earth's fate, there's probably no more instructive case study than the Florida Everglades. When European explorers first arrived there in the mid-1800s, they discovered a lush, tropical wilderness with dense sawgrass, marshes, mangrove forests, lakes, and tree islands. By the early 20th century, developers and politicians had begun building a series of canals and dikes to siphon off the region's water. They succeeded in creating an agricultural and real estate boom, and to some degree, they offset floods and droughts. But the ecological cost was exorbitant. Today, half of the Everglades' wetlands have been lost, its water is polluted by runoff from farms, and much of its wildlife, including Florida panthers and many wading birds such as wood storks, are hanging on by a thread.

Yet there has been a renewed sense of hope in the Everglades since 2001, when the state of Florida and the federal government approved a comprehensive $7.8 billion restoration plan, the biggest recovery of its kind in history. During the next four decades, ecologists and engineers will work to undo years of ecological damage by redirecting water back into the Everglades' dried-up marshes. "The Everglades are a test," says Joe Podger, an environmentalist. "If we pass, we get to keep the planet."

In fact, as this comprehensive series on environmental issues shows, humankind faces a host of tests that will determine whether we get to keep the planet. The world's crises—air and water pollution, the extinction of species, and climate change—are worsening by the day. The solutions—and there are many practical ones—all demand an extreme sense of urgency. E. O. Wilson, the noted Harvard zoologist, contends that "the world environment is changing so fast that there is a window of opportunity that will close in as little time as the next two or three decades." While Wilson's main concern is the rapid loss of biodiversity, he could have just as easily been discussing climate change or wetlands destruction.

The Earth is suffering the most massive extinction of species since the die-off of dinosaurs 65 million years ago. "If we continue at the current rate of deforestation and destruction of major ecosystems like rain forests and coral reefs, where most of the biodiversity is concentrated," Wilson says, "we will surely lose more than half of all the species of plants and animals on Earth by the end of the 21st century."

Many conservationists still mourn the loss of the passenger pigeon, which, as recently as the late 1800s, flew in miles-long flocks so dense they blocked the sun, turning noontime into nighttime. By 1914, target shooters and market hunters had reduced the species to a single individual, Martha, who lived at the Cincinnati Zoo until, as Peter Matthiessen wrote in *Wildlife in America,* "she blinked for the last time." Despite U.S. laws in place to avert other species from going the way of the passenger pigeon, the latest news is still alarming. In its 2004 State of the Birds report, Audubon noted that 70% of grassland bird species and 36% of shrubland bird species are suffering significant declines. Like the proverbial canary in the coalmine, birds serve as indicators, sounding the alarm about impending threats to environmental and human health.

Besides being an unmitigated moral tragedy, the disappearance of species has profound practical implications. Ninety percent of the world's food production now comes from about a dozen species of plants and eight species of livestock. Geneticists rely on wild populations to replenish varieties of domestic corn, wheat, and other crops, and to boost yields and resistance to disease. "Nature is a natural pharmacopoeia, and new drugs and medicines are being discovered in the wild all the time," wrote Niles Eldredge of the American Museum of Natural History, a noted author on the subject of extinction. "Aspirin comes from the bark of willow trees. Penicillin comes from a mold, a type of fungus." Furthermore, having a wide array of plants and animals improves a region's capacity to cleanse water, enrich soil, maintain stable climates, and produce the oxygen we breathe.

Today, the quality of the air we breathe and the water we drink does not augur well for our future health and well-being. Many people assume that the passage of the Clean Air Act in 1970 ushered in a new age. But the American Lung Association reports that 159 million Americans—55% of the population—are exposed to "unhealthy levels of air pollution." Meanwhile, the American Heart Association warns of a direct link between exposure to air pollution and heart disease and strokes. While it's true that U.S. waters are cleaner than they were three decades ago, data from the Environmental Protection Agency (EPA) shows that almost half of U.S. coastal waters fail to meet water-quality standards because they cannot support fishing or swimming. Each year, contaminated tap water makes as many as 7 million Americans sick. The chief cause is "non-point pollution," runoff that includes fertilizers and pesticides from farms and backyards as well as oil and chemical spills. On a global level, more than a billion people lack access to clean water; according to the United Nations, five times that number die each year from malaria and other illnesses associated with unsafe water.

Of all the Earth's critical environmental problems, one trumps the rest: climate change. Carol Browner, the EPA's chief from 1993 through 2001 (the longest term in the agency's history), calls climate change "the greatest environmental health problem the world has ever seen." Industry and people are spewing carbon dioxide from smokestacks and the tailpipes of their cars into the atmosphere, where a buildup of gases, acting like the glass in a greenhouse, traps the sun's heat. The 1990s was the warmest decade in more than a century, and 1998 saw the highest global temperatures ever. In an article about global climate change in the December 2003 issue of *Audubon*, David Malakoff wrote, "Among the possible consequences: rising sea levels that cause coastal communities to sink beneath the waves like a modern Atlantis, crop failures of biblical proportions, and once-rare killer storms that start to appear with alarming regularity."

Yet for all the doom and gloom, scientists and environmentalists hold out hope. When Russia recently ratified the Kyoto Protocol, it meant that virtually all of the world's industrialized nations—the United States, which has refused to sign, is a notable exception—have committed to cutting greenhouse gases. As Kyoto and other international agreements go into effect, a market is developing for cap-and-trade systems for carbon dioxide. In this country, two dozen big corporations, including British Petroleum, are cutting emissions. At least 28 American states have adopted their own policies. California, for example, has passed global warming legislation aimed at curbing emissions from new cars. Governor Arnold Schwarzenegger has also backed regulations requiring automakers to slash the amount of greenhouse gases they cause by up to 30% by 2016, setting a precedent for other states.

As Washington pushes a business-friendly agenda, states are filling in the policy vacuum in other areas, as well. California and New York are developing laws to preserve wetlands, which filter pollutants, prevent floods, and provide habitat for endangered wildlife.

By taking matters into their own hands, states and foreign countries will ultimately force Washington's. What industry especially abhors is a crazy quilt of varying rules. After all, it makes little sense for a company to invest a billion dollars in a power plant only to find out later that it has to spend even more to comply with another state's stricter emissions standards. Ford chairman and chief executive William Ford has lashed out at the states' "patchwork" approach because he and "other manufacturers will have a hard time responding." Further, he wrote in a letter to his company's top managers, "the prospect of 50 different requirements in 50 different states would be nothing short of chaos." The type of fears Ford expresses are precisely the reason federal laws protecting clean air and water came into being.

Governments must take the lead, but ecologically conscious consumers wield enormous influence, too. Over the past four decades, the annual use of pesticides has more than doubled, from 215 million pounds to 511 million pounds. Each year, these poisons cause $10 billion worth of damage to the environment and kill 72 million birds. The good news is that the demand for organic products is revolutionizing agriculture, in part by creating a market for natural alternatives for pest control. Some industry experts predict that by 2007 the organic industry will almost quadruple, to more than $30 billion.

E. O. Wilson touts "shade-grown" coffee as one of many "personal habitats that, if moderated only in this country, could contribute significantly to saving endangered species." In the mountains of Mexico and Central America, coffee grown beneath a dense forest canopy rather than in cleared fields helps provide refuge for dozens of wintering North American migratory bird species, from western tanagers to Baltimore orioles.

With conservation such a huge part of Americans' daily routine, recycling has become as ingrained a civic duty as obeying traffic lights. Californians, for their part, have cut their energy consumption by 10% each year since the state's 2001 energy crisis. "Poll after poll shows that about two-thirds of the American public—Democrat and Republican, urban and rural—consider environmental progress crucial," writes Carl Pope, director of the Sierra Club, in his recent book, *Strategic Ignorance*. "Clean air, clean water, wilderness preservation—these are such bedrock values that many polling respondents find it hard to believe that any politician would oppose them."

Terrorism and the economy clearly dwarfed all other issues in the 2004 presidential election. Even so, voters approved 120 out of 161 state and local conservation funding measures nationwide, worth a total of $3.25 billion. Anti-environment votes in the U.S. Congress and proposals floated by the like-minded Bush administration should not obscure the salient fact that so far there have been no changes to the major environmental laws. The potential for political fallout is too great.

The United States' legacy of preserving its natural heritage is the envy of the world. Our national park system alone draws more than 300 million visitors each year. Less well known is the 103-year-old national wildlife refuge system you'll learn about in this series. Its unique mission is to safeguard the nation's wild animals and plants on 540 refuges, protecting 700 species of birds and an equal number of other vertebrates; 282 of these species are either threatened or endangered. One of the many species particularly dependent on the invaluable habitat refuges afford is the bald eagle. Such safe havens, combined with the banning of the insecticide DDT and enforcement of the

Endangered Species Act, have led to the bald eagle's remarkable recovery, from a low of 500 breeding pairs in 1963 to 7,600 today. In fact, this bird, the national symbol of the United States, is about be removed from the endangered species list and downgraded to a less threatened status under the CITES, the Convention on International Trade in Endangered Species.

This vital treaty, upheld by the United States and 165 other participating nations (and detailed in this series), underscores the worldwide will to safeguard much of the Earth's magnificent wildlife. Since going into effect in 1975, CITES has helped enact plans to save tigers, chimpanzees, and African elephants. These species and many others continue to face dire threats from everything from poaching to deforestation. At the same time, political progress is still being made. Organizations like the World Wildlife Fund work tirelessly to save these species from extinction because so many millions of people care. China, for example, the most populous nation on Earth, is so concerned about its giant pandas that it has implemented an ambitious captive breeding program. That program's success, along with government measures prohibiting logging throughout the panda's range, may actually enable the remaining population of 1,600 pandas to hold its own—and perhaps grow. "For the People's Republic of China, pressure intensified as its internationally popular icon edged closer to extinction," wrote Gerry Ellis in a recent issue of *National Wildlife.* "The giant panda was not only a poster child for endangered species, it was a symbol of our willingness to ensure nature's place on Earth."

Whether people take a spiritual path to conservation or a pragmatic one, they ultimately arrive at the same destination. The sight of a bald eagle soaring across the horizon reassures us about nature's resilience, even as the clean air and water we both need to survive becomes less of a certainty. "The conservation of our natural resources and their proper use constitute the fundamental problem which underlies almost every other problem of our national life," President Theodore Roosevelt told Congress at the dawn of the conservation movement a century ago. His words ring truer today than ever.

Introduction: "Why Should We Care?"

Our nation's air and water are cleaner today than they were 30 years ago. After a century of filling and destroying over half of our wetlands, we now protect many of them. But the Earth is getting warmer, habitats are being lost to development and logging, and humans are using more water than ever before. Increased use of water can leave rivers, lakes, and wetlands without enough water to support the native plant and animal life. Such changes are causing plants and animals to go extinct at an increased rate. It is no longer a question of losing just the dodo birds or the passenger pigeons, argues David Quammen, author of *Song of the Dodo*: "Within a few decades, if present trends continue, we'll be losing *a lot* of everything."[1]

In the 1980s, E. O. Wilson, a Harvard biologist and Pulitzer Prize–winning author, helped bring the term *biodiversity* into public discussions about conservation. *Biodiversity*, short for "biological diversity," refers to the levels of organization for living things. Living organisms are divided and categorized into ecosystems (such as rain forests or oceans), by species (such as mountain gorillas), and by genetics (the genes responsible for inherited traits).

Wilson has predicted that if we continue to destroy habitats and pollute the Earth at the current rate, in 50 years, we could lose 30 to 50% of the planet's species to extinction. In his 1992 book, *The Diversity of Life*, Wilson asks: "Why should we care?"[2] His long list of answers to this question includes: the potential loss of vast amounts of scientific information that would enable the development of new crops, products, and medicines and the potential loss of the vast economic and environmental benefits of healthy ecosystems. He argues that since we have only a vague idea (even with our advanced scientific methods) of how ecosystems really work, it would be "reckless" to suppose that destroying species indefinitely will not threaten us all in ways we may not even understand.

THE BOOKS IN THE SERIES

In looking at environmental issues, it quickly becomes clear that, as naturalist John Muir once said, "When we try to pick out anything by itself, we find it hitched to everything else in the Universe."[3] For example, air pollution in one state or in one country can affect not only air quality in another place, but also land and water quality. Soil particles from degraded African lands can blow across the ocean and cause damage to far-off coral reefs.

The six books in this series address a variety of environmental issues: conservation, wildlife protection, water pollution, air quality, climate change, and environmental policy. None of these can be viewed as a separate issue. Air quality impacts climate change, wildlife, and water quality. Conservation initiatives directly affect water and air quality, climate change, and wildlife protection. Endangered species are touched by each of these issues. And finally, environmental policy issues serve as important tools in addressing all the other environmental problems that face us.

You can use the burning of coal as an example to look at how a single activity directly "hitches" to a variety of environmental issues. Humans have been burning coal as a fuel for hundreds of years. The mining of coal can leave the land stripped of vegetation, which erodes the soil. Soil erosion contributes to particulates in the air and water quality problems. Mining coal can also leave piles of acidic tailings that degrade habitats and pollute water. Burning any fossil fuel—coal, gas, or oil—releases large amounts of carbon dioxide into the atmosphere. Carbon dioxide is considered a major "greenhouse gas" that contributes to global warming—the gradual increase in the Earth's temperature over time. In addition, coal burning adds sulfur dioxide to the air, which contributes to the formation of acid rain—precipitation that is abnormally acidic. This acid rain can kill forests and leave lakes too acidic to support life. Technology continues to present ways to minimize the pollution that results from extracting and burning fossil fuels. Clean air and climate change policies guide states and industries toward implementing various strategies and technologies for a cleaner coal industry.

Each of the six books in this series—ENVIRONMENTAL ISSUES—introduces the significant points that relate to the specific topic and explains its relationship to other environmental concerns.

Book One: *Air Quality*

Problems of air pollution can be traced back to the time when humans first started to burn coal. *Air Quality* looks at today's challenges in fighting to keep our air clean and safe. The book includes discussions of air pollution sources—car and truck emissions, diesel engines, and many industries. It also discusses their effects on our health and the environment.

The Environmental Protection Agency (EPA) has reported that more than 150 million Americans live in areas that have unhealthy levels of some type of air pollution.[4] Today, more than 20 million Americans, over 6 million of whom are children, suffer from asthma believed to be triggered by pollutants in the air.[5]

In 1970, Congress passed the Clean Air Act, putting in place an ambitious set of regulations to address air pollution concerns. The EPA has identified and set standards for six common air pollutants: ground-level ozone, nitrogen oxides, particulate matter, sulfur dioxide, carbon monoxide, and lead.

The EPA has also been developing the Clean Air Rules of 2004, national standards aimed at improving the country's air quality by specifically addressing the many sources of contaminants. However, many conservation organizations and even some states have concerns over what appears to be an attempt to weaken different sections of the 1990 version of the Clean Air Act. The government's environmental protection efforts take on increasing importance because air pollution degrades land and water, contributes to global warming, and affects the health of plants and animals, including humans.

Book Two: *Climate Change*

Part of science is observing patterns, and scientists have observed a global rise in temperature. *Climate Change* discusses the sources and effects of global warming. Scientists attribute this accelerated change to human activities such as the burning of fossil fuels that emit greenhouse gases (GHG).[6] Since the 1700s, we have been cutting down the trees that help remove carbon dioxide from the atmosphere, and have increased the amount of coal, gas, and oil we burn, all of which add carbon dioxide to the atmosphere. Science tells us that these human activities have caused greenhouse gases—carbon dioxide (CO_2), methane (CH_4), nitrous oxide (N_2O), hydrofluorocarbons (HFCs), perfluorocarbons (PFCs), and sulfur hexafluoride (SF_6)—to accumulate in the atmosphere.[7]

If the warming patterns continue, scientists warn of more negative environmental changes. The effects of climate change, or global warming, can be seen all over the world. Thousands of scientists are predicting rising sea levels, disturbances in patterns of rainfall and regional weather, and changes in ranges and reproductive cycles of plants and animals. Climate change is already having some effects on certain plant and animal species.[8]

Many countries and some American states are already working together and with industries to reduce the emissions of greenhouse gases. Climate change is an issue that clearly fits noted scientist Rene Dubois's advice: "Think globally, act locally."

Book Three: *Conservation*

Conservation considers the issues that affect our world's vast array of living creatures and the land, water, and air they need to survive.

One of the first people in the United States to put the political spotlight on conservation ideas was President Theodore Roosevelt. In the early 1900s, he formulated policies and created programs that addressed his belief that: "The nation behaves well if it treats the natural resources as assets which it must turn over to the next generation increased, and not impaired, in value."[9] In the 1960s, biologist Rachel Carson's book, *Silent Spring*, brought conservation issues into the public eye. People began to see that polluted land, water, and air affected their health. The 1970s brought the creation of the United States Environmental Protection Agency (EPA) and passage of many federal and state rules and regulations to protect the quality of our environment and our health.

Some 80 years after Theodore Roosevelt established the first National Wildlife Refuge in 1903, Harvard biologist

E. O. Wilson brought public awareness of conservation issues to a new level. He warned:

> . . . the worst thing that will probably happen—in fact is already well underway—is not energy depletion, economic collapse, conventional war, or even the expansion of totalitarian governments. As terrible as these catastrophes would be for us, they can be repaired within a few generations. The one process now ongoing that will take million of years to correct is the loss of genetic species diversity by the destruction of natural habitats. This is the folly our descendants are least likely to forgive us.[10]

To heed Wilson's warning means we must strive to protect species-rich habitats, or "hotspots," such as tropical rain forests and coral reefs. It means dealing with conservation concerns like soil erosion and pollution of fresh water and of the oceans. It means protecting sea and land habitats from the overexploitation of resources. And it means getting people involved on all levels—from national and international government agencies, to private conservation organizations, to the individual person who recycles or volunteers to listen for the sounds of frogs in the spring.

Book Four: *Environmental Policy*

One approach to solving environmental problems is to develop regulations and standards of safety. Just as there are rules for living in a community or for driving on a road, there are environmental regulations and policies that work toward protecting our health and our lands. *Environmental Policy* discusses the regulations and programs that have been crafted to address environmental issues at all levels—global, national, state, and local.

Today, as our resources become increasingly limited, we witness heated debates about how to use our public lands and how to protect the quality of our air and water. Should we allow drilling in the Arctic National Wildlife Refuge? Should we protect more marine areas? Should we more closely regulate the emissions of vehicles, ships, and industries? These policy issues, and many more, continue to make news on a daily basis.

In addition, environmental policy has taken a place on the international front. Hundreds of countries are working together in a variety of ways to address such issues as global warming, air pollution, water pollution and supply, land preservation, and the protection of endangered species. One question the United States continues to debate is whether to sign the 1997 Kyoto Protocol, the international agreement designed to decrease the emissions of greenhouse gases.

Many of the policy tools for protecting our environment are already in place. It remains a question how they will be used—and whether they will be put into action in time to save our natural resources and ourselves.

Book Five: *Water Pollution*

Pollution can affect water everywhere. Pollution in lakes and rivers is easily seen. But water that is out of our plain view can also be polluted with substances such as toxic chemicals, fertilizers, pesticides, oils, and gasoline. *Water Pollution* considers issues of concern to our surface waters, our groundwater, and our oceans.

In the early 1970s, about three-quarters of the water in the United States was considered unsafe for swimming and fishing. When Lake Erie was declared "dead" from pollution and a river feeding it actually caught on fire, people decided that the national government had to take a stronger role in protecting our resources. In 1972, Congress passed the Clean Water Act, a law whose objective "is to restore and maintain the chemical, physical, and biological integrity of the Nation's waters."[11] Today, over 30 years later, many lakes and rivers have been restored to health. Still, an estimated 40% of our waters are still unsafe to swim in or fish.

Less than 1% of the available water on the planet is fresh water. As the world's population grows, our demand for drinking and irrigation water increases. Therefore, the quantity of available water has become a major global issue. As Sandra Postel, a leading authority on international freshwater issues, says, "Water scarcity is now the single biggest threat to global food production."[12] Because there are many competing demands for water, including the needs of habitats, water pollution continues to become an even more serious problem each year.

Book Six: *Wildlife Protection*

For many years, the word *wildlife* meant only the animals that people hunted for food or for sport. It was not until 1986 that the Oxford English Dictionary defined *wildlife* as "the native fauna and flora of a particular region."[13] *Wildlife Protection* looks at overexploitation—for example, overfishing or collecting plants and animals for illegal trade—and habitat loss. Habitat loss can be the result of development, logging, pollution, water diverted for human use, air pollution, and climate change.

Also discussed are various approaches to wildlife protection. Since protection of wildlife is an issue of global concern, it is addressed here on international as well as on national and local levels. Topics include voluntary international organizations such as the International Whaling Commission and the CITES agreements on trade in endangered species. In the United States, the Endangered Species Act provides legal protection for more than 1,200 different plant and animal species. Another approach to wildlife protection includes developing partnerships among conservation organizations, governments, and local people to foster economic incentives to protect wildlife.

CONSERVATION IN THE UNITED STATES

Those who first lived on this land, the Native American peoples, believed in general that land was held in common, not to be individually owned, fenced, or tamed. The white settlers from Europe had very different views of land. Some believed the New World was a Garden of Eden. It was a land of opportunity for them, but it was also a land to be controlled and subdued. Ideas on how to treat the land often followed those of European thinkers like John Locke, who believed that "Land that is left wholly to nature is called, as indeed it is, waste." [14]

The 1800s brought another way of approaching the land. Thinkers such as Ralph Waldo Emerson, John Muir, and Henry David Thoreau celebrated our human connection with nature. By the end of the 1800s, some scientists and policymakers were noticing the damage humans have caused to the land. Leading public officials preached stewardship and wise use of our country's resources. In 1873, Yellowstone National Park was set up. In 1903, the first National Wildlife Refuge was established.

However, most of the government practices until the middle of the 20th century favored unregulated development and use of the land's resources. Forests were clear cut, rivers were dammed, wetlands were filled to create farmland, and factories were allowed to dump their untreated waste into rivers and lakes.

In 1949, a forester and ecologist named Aldo Leopold revived the concept of preserving land for its own sake. But there was now a biological, or scientific, reason for conservation, not just a spiritual one. Leopold declared: "All ethics rest upon a single premise: that the individual is a member of a community of interdependent parts. . . . A thing is right when it tends to preserve the integrity and stability and beauty of the biotic community. It is wrong when it tends otherwise." [15]

The fiery vision of these conservationists helped shape a more far-reaching movement that began in the 1960s. Many credit Rachel Carson's eloquent and accessible writings, such as her 1962 book *Silent Spring*, with bringing environmental issues into people's everyday language. When the Cuyahoga River in Ohio caught fire in 1969 because it was so polluted, it captured the public attention. Conservation was no longer just about protecting land that many people would never even see, it was about protecting human health. The condition of the environment had become personal.

In response to the public outcry about water and air pollution, the 1970s saw the establishment of the EPA. Important legislation to protect the air and water was passed. National standards for a cleaner environment were set and programs were established to help achieve the ambitious goals. Conservation organizations grew from what had started as exclusive white men's hunting clubs to interest groups with a broad membership base. People came together to demand changes that would afford more protection to the environment and to their health.

Since the 1960s, some presidential administrations have sought to strengthen environmental protection and to protect more land and national treasures. For example, in 1980, President Jimmy Carter signed an act that doubled the amount of protected land in Alaska and renamed it the Arctic National Wildlife Refuge. Other administrations, like those of President Ronald Reagan, sought to dismantle many earlier environmental protection initiatives.

The environmental movement, or environmentalism, is not one single, homogeneous cause. The agencies, individuals, and organizations that work toward protecting the environment vary as widely as the habitats and places they seek to protect. There are individuals who begin grass-roots efforts—people like Lois Marie Gibbs, a former resident of the polluted area of Love Canal, New York, who founded the Center for Health, Environment and Justice. There are conservation organizations, like The Nature Conservancy, the World Wildlife Fund (WWF), and Conservation International, that sponsor programs to preserve and protect habitats. There are groups that specialize in monitoring public policy and legislation—for example, the Natural Resources Defense Council and Environmental Defense. In addition, there are organizations like the Audubon Society and the National Wildlife Federation whose focus is on public education about environmental issues. Perhaps from this diversity, just like there exists in a healthy ecosystem, will come the strength and vision environmentalism needs to deal with the continuing issues of the 21st century.

INTERNATIONAL CONSERVATION EFFORTS

In his book *Biodiversity*, E. O. Wilson cautions that biological diversity must be taken seriously as a global resource for three reasons. First, human population growth is accelerating the degrading of the environment, especially in tropical countries. Second, science continues to discover new uses for biological diversity—uses that can benefit human health and protect the environment. And third, much biodiversity is being lost through extinction, much of it in the tropics. As Wilson states, "We must hurry to acquire the knowledge on which a wise policy of conservation and development can be based for centuries to come."[16]

People organize themselves within boundaries and borders. But oceans, rivers, air, and wildlife do not follow such rules. Pollution or overfishing in one part of an ocean can easily degrade the quality of another country's resources. If one country diverts a river, it can destroy another country's wetlands or water resources. When Wilson cautions us that we must hurry to develop a wise conservation policy, he means a policy that will protect resources all over the world.

To accomplish this will require countries to work together on critical global issues: preserving biodiversity, reducing global warming, decreasing air pollution, and protecting the oceans. There are many important international efforts already going on to protect the resources of our planet. Some efforts are regulatory, while others are being pursued by nongovernmental organizations or private conservation groups.

Countries volunteering to cooperate to protect resources is not a new idea. In 1946, a group of countries established the International Whaling Commission (IWC). They recognized that unregulated whaling around the world had led to severe declines in the world's whale populations. In 1986, the IWC declared a moratorium on whaling, which is still in effect, until the populations have recovered.[17] Another example of international cooperation occurred in 1987 when various countries signed the Montreal Protocol to reduce the emissions of ozone-depleting gases. It has been a huge success, and perhaps has served as a model for other international efforts, like the 1997 Kyoto Protocol, to limit emissions of greenhouse gases.

Yet another example of international environmental cooperation is the CITES agreement (the Convention on International Trade in Endangered Species of Wild Fauna and Flora), a legally binding agreement to ensure that the international trade of plants and animals does not threaten the species' survival. CITES went into force in 1975 after 80 countries agreed to the terms. Today, it has grown to include more than 160 countries. This make CITES among the largest conservation agreements in existence.[18]

Another show of international conservation efforts are governments developing economic incentives for local conservation. For example, in 1996, the International Monetary Fund (IMF) and the World Wildlife Fund (WWF) established a program to relieve poor countries of debt. More than 40 countries have benefited by agreeing to direct some of their savings toward environmental programs in the "Debt-for-Nature" swap programs.[19]

It is worth our time to consider the thoughts of two American conservationists and what role we, as individuals, can play in conserving and protecting our world. E. O. Wilson has told us that "Biological Diversity—'biodiversity' in the new parlance—is the key to the maintenance of the world as we know it."[20] Aldo Leopold, the forester who gave Americans the idea of creating a "land ethic," wrote in 1949 that: "Having to squeeze the last drop of utility out of the land has the same desperate finality as having to chop up the furniture to keep warm."[21] All of us have the ability to take part in the struggle to protect our environment and to save our endangered Earth.

ENDNOTES

1   Quammen, David. *Song of the Dodo*. New York: Scribner, 1996, p. 607.

2   Wilson, E. O. *Diversity of Life*. Cambridge, MA: Harvard University Press, 1992, p. 346.

3   Muir, John. *My First Summer in the Sierra*. San Francisco: Sierra Club Books, 1988, p. 110.

4   Press Release. *EPA Newsroom: EPA Issues Designations on Ozone Health Standards.* April 15, 2004. Available online at *http://www.epa.gov/newsroom/.*

5   The Environmental Protection Agency. EPA Newsroom. *May is Allergy Awareness Month.* May 2004. Available online at *http://www.epa.gov/newsroom/allergy_month.htm.*

6   Intergovernmental Panel on Climate Change (IPCC). Third Annual Report, 2001.

7   Turco, Richard P. *Earth Under Siege: From Air Pollution to Global Change.* New York: Oxford University Press, 2002, p. 387.

8   Intergovernmental Panel on Climate Change. *Technical Report V: Climate Change and Biodiversity.* 2002. Full report available online at *http://www.ipcc.ch/pub/tpbiodiv.pdf.*

9   "Roosevelt Quotes." American Museum of Natural History. Available online at *http://www.amnh.org/common/faq/quotes.html.*

10 Wilson, E. O. *Biophilia.* Cambridge, MA: Harvard University Press, 1986, pp. 10–11.

11 Federal Water Pollution Control Act. As amended November 27, 2002. Section 101 (a).

12 Postel, Sandra. *Pillars of Sand.* New York: W. W. Norton & Company, Inc., 1999. p. 6.

13 Hunter, Malcolm L. *Wildlife, Forests, and Forestry: Principles of Managing Forest for Biological Diversity.* Englewood Cliffs, NJ: Prentice-Hall, 1990, p. 4.

14 Dowie, Mark. *Losing Ground: American Environmentalism at the Close of the Twentieth Century.* Cambridge, MA: MIT Press, 1995, p. 113.

15 Leopold, Aldo. *A Sand County Almanac.* New York: Oxford University Press, 1949.

16 Wilson, E. O., ed. *Biodiversity.* Washington, D.C.: National Academies Press, 1988, p. 3.

17 International Whaling Commission Information 2004. Available online at *http://www.iwcoffice.org/commission/iwcmain.htm.*

18 *Discover CITES: What is CITES?* Fact sheet 2004. Available online at *http://www.cites.org/eng/disc/what.shtml.*

19 *Madagascar's Experience with Swapping Debt for the Environment.* World Wildlife Fund Report, 2003. Available online at *http://www.conservationfinance.org/WPC/WPC_documents/ Apps_11_Moye_Paddack_v2.pdf.*

20 Wilson, *Diversity of Life,* p. 15.

21 Leopold.

Section A
Water Pollution Issues and Challenges

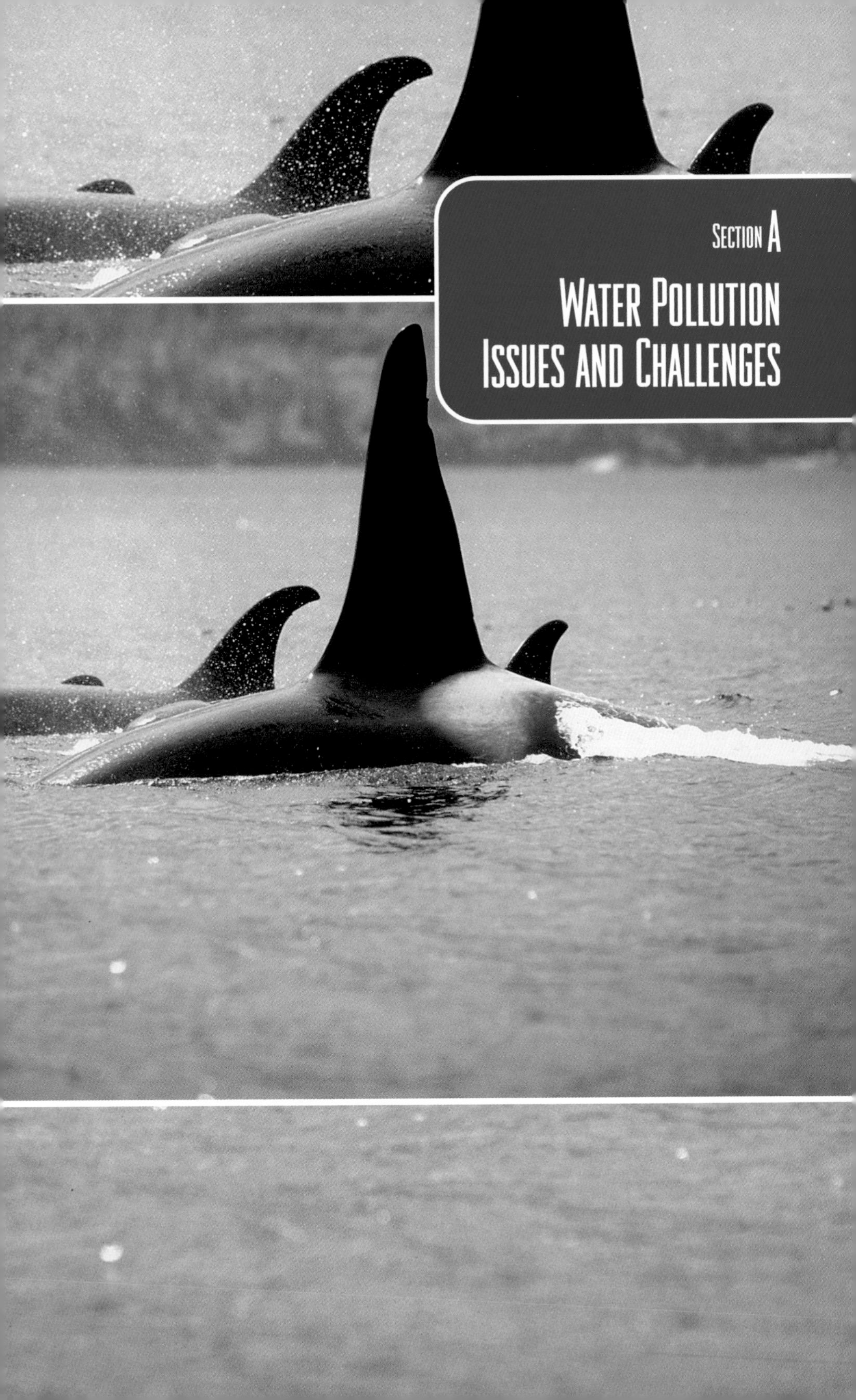

Why Is Polluted Water an Issue in the United States and All Over the World?

What does clean water have to do with you? Well, to start with, your body is about three-quarters water. The food you eat is grown with water. The oxygen you breathe is generated by plants that use water. These same plants clean the air you breathe. And some things you might like to do require clean water—swimming, cooking, and showering are examples. If all water were polluted, living things, including you, would get sick. Somewhere in the world, a child dies every eight seconds from a disease related to contaminated water.

There are many sources of water pollution: direct discharges by industry, sewage, leaking underground storage tanks, and landfills are some point sources. But a major source of pollution comes from sources that are harder to pinpoint, and, therefore, harder to control. These nonpoint sources, as they are called, include run-off that contains fertilizers, herbicides, and pesticides from agricultural lands and drainage from paved surfaces and storm drains. Another source is not quite as obvious, but is deadly serious. Pollution from the air falls to the water as acid rain, killing lakes and causing harmful substances such as mercury to accumulate in fish.

Polluted water is not a problem in the United States alone. The United Nations (UN) declared 2003 the international year of fresh water, as an effort to rally international and regional support for water issues. The first article that follows, *The Quest for Clean Water*, identifies some global water issues and examines what is being done to address them. The second excerpt, from a 2003 World Bank and World Wildlife Fund report, *Running Pure: The Importance of Forest Protected Areas to Drinking Water*, provides some background on the global issues.

—The Editor

The Quest for Clean Water
by Joseph Orlins and Anner Wehrly

As water pollution threatens our health and environment, we need to implement an expanding array of techniques for its assessment, prevention, and remediation.

In the 1890s, entrepreneur William Love sought to establish a model industrial community in the La Salle district of Niagara Falls, New York. The plan included building a canal that tapped water from the Niagara River for a navigable waterway and a hydroelectric power plant. Although work on the canal was begun, a nationwide economic depression and other factors forced abandonment of the project.

By 1920, the land adjacent to the canal was sold and used as a landfill for municipal and industrial wastes. Later purchased by Hooker Chemicals and Plastics Corp., the landfill became a dumping ground for nearly 21,000 tons of mixed chemical wastes before being closed and covered over in the early 1950s. Shortly thereafter, the property was acquired by the Niagara Falls Board of Education, and schools and residences were built on and around the site.

In the ensuing decades, groundwater levels in the area rose, parts of the landfill subsided, large metal drums of waste were uncovered, and toxic chemicals oozed out. All this led to the contamination of surface waters, oily residues in residential basements, corrosion of sump pumps, and noxious odors. Residents began to question if these problems were at the root of an apparent prevalence of birth defects and miscarriages in the neighborhood.

Eventually, in 1978, the area was declared unsafe by the New York State Department of Health, and President Jimmy Carter approved emergency federal assistance. The school located on the landfill site was closed and nearby houses were condemned. State and federal agencies worked together to relocate hundreds of residents and contain or destroy the chemical wastes.

That was the bitter story of Love Canal. Although not the worst environmental disaster in U.S. history, it illustrates the tragic consequences of water pollution.

WATER QUALITY STANDARDS

In addition to toxic chemical wastes, water pollutants occur in many other forms, including pathogenic microbes (harmful bacteria and viruses), excess fertilizers (containing compounds of phosphorus and nitrogen), and trash floating on streams, lakes, and beaches. Water pollution can also take the form of sediment eroded from stream banks, large blooms of algae, low levels of dissolved oxygen, or abnormally high temperatures (from the discharge of coolant water at power plants).

The United States has seen a growing concern about water pollution since the middle of the twentieth century, as the public recognized that pollutants were adversely affecting human health and rendering lakes unswimmable, streams unfishable, and rivers flammable. In response, in 1972, Congress passed the Federal Water Pollution Control Act Amendments, later modified and referred to as the Clean Water Act. Its purpose was to "restore and maintain the chemical, physical, and biological integrity of the nation's waters."

The Clean Water Act set the ambitious national goal of completely eliminating the discharge of pollutants into navigable waters by 1985, as well as the interim goal of making water clean enough to sustain fish and wildlife, while being safe for swimming and boating. To achieve these goals, certain standards for water quality were established.

The "designated uses" of every body of water subject to the act must first be identified. Is it a source for drinking water? Is it used for recreation, such as swimming? Does it supply agriculture or industry? Is it a significant habitat for fish and other aquatic life? Thereafter, the water must be tested for pollutants. If it fails to meet the minimum standards for its designated uses, then steps must be taken to limit pollutants entering it, so that it becomes suitable for those uses.

On the global level, the fundamental importance of clean water has come into the spotlight. In November 2002, the UN Committee on Economic, Cultural and Social Rights declared access to clean water a human right. Moreover, the United Nations has designated 2003 to be the International Year of

Freshwater, with the aim of encouraging sustainable use of freshwater and integrated water resources management.

HERE, THERE, AND EVERYWHERE

Implementing the Clean Water Act requires clarifying the sources of pollutants. They are divided into two groups: "point sources" and "nonpoint sources." Point sources correspond to discrete, identifiable locations from which pollutants are emitted. They include factories, wastewater treatment plants, landfills, and underground storage tanks. Water pollution that originates at point sources is usually what is associated with headline-grabbing stories such as those about Love Canal.

Nonpoint sources of pollution are diffuse and therefore harder to control. For instance, rain washes oil, grease, and solid pollutants from streets and parking lots into storm drains that carry them into bays and rivers. Likewise, irrigation and rainwater leach fertilizers, herbicides, and insecticides from farms and lawns and into streams and lakes.

The direct discharge of wastes from point sources into lakes, rivers, and streams is regulated by a permit program known as the National Pollutant Discharge Elimination System (NPDES). This program, established through the Clean Water Act, is administered by the Environmental Protection Agency (EPA) and authorized states. By regulating the wastes discharged, NPDES has helped reduce point-source pollution dramatically. On the other hand, water pollution in the United States is now mainly from nonpoint sources, as reported by the EPA.

In 1991, the U.S. Geological Survey (USGS, part of the Department of the Interior) began a systematic, long-term program to monitor watersheds. The National Water-Quality Assessment Program (NAWQA), established to help manage surface and groundwater supplies, has involved the collection and analysis of water quality data in over 50 major river basins and aquifer systems in nearly all 50 states.

The program has encompassed three principal categories of investigation: (1) the current conditions of surface water and groundwater; (2) changes in those conditions over time; and (3) major factors—such as climate, geography, and land use—that affect water quality. For each of these categories, the water and sediment have been tested for such pollutants as pesticides, plant nutrients, volatile organic compounds, and heavy metals.

The NAWQA findings were disturbing. Water quality is most affected in watersheds with highest population density and urban development. In agricultural areas, 95 percent of tested streams and 60 percent of shallow wells contained herbicides, insecticides, or both. In urban areas, 99 percent of tested streams and 50 percent of shallow wells had herbicides, especially those used on lawns and golf courses. Insecticides were found more frequently in urban streams than in agricultural ones.

The study also found large amounts of plant nutrients in water supplies. For instance, 80 percent of agricultural streams and 70 percent of urban streams were found to contain phosphorus at concentrations that exceeded EPA guidelines.

Moreover, in agricultural areas, one out of five well-water samples had nitrate concentrations higher than EPA standards for drinking water. Nitrate contamination can result from nitrogen fertilizers or material from defective septic systems leaching into the groundwater, or it may reflect defects in the wells.

EFFECTS OF POLLUTION

According to the UN World Water Assessment Programme, about 2.3 billion people suffer from diseases associated with polluted water, and more than 5 million people die from these illnesses each year. Dysentery, typhoid, cholera, and hepatitis A are some of the ailments that result from ingesting water contaminated with harmful microbes. Other illnesses—such as malaria, filariasis, yellow fever, and sleeping sickness—are transmitted by vector organisms (such as mosquitoes and tsetse flies) that breed in or live near stagnant, unclean water.

A number of chemical contaminants—including DDT, dioxins, polychlorinated biphenyls (PCBs), and heavy metals—

are associated with conditions ranging from skin rashes to various cancers and birth defects. Excess nitrate in an infant's drinking water can lead to the "blue baby syndrome" (methemoglobinemia)—a condition in which the child's digestive system cannot process the nitrate, diminishing the blood's ability to carry adequate concentrations of oxygen.

Besides affecting human health, water pollution has adverse effects on ecosystems. For instance, while moderate amounts of nutrients in surface water are generally not problematic, large quantities of phosphorus and nitrogen compounds can lead to excessive growth of algae and other nuisance species. Known as *eutrophication,* this phenomenon reduces the penetration of sunlight through the water; when the plants die and decompose, the body of water is left with odors, bad taste, and reduced levels of dissolved oxygen.

Low levels of dissolved oxygen can kill fish and shellfish. In addition, aquatic weeds can interfere with recreational activities (such as boating and swimming) and can clog intake by industry and municipal systems.

Some pollutants settle to the bottom of streams, lakes, and harbors, where they may remain for many years. For instance, although DDT and PCBs were banned years ago, they are still found in sediments in many urban and rural streams. They occur at levels harmful to wildlife at more than two-thirds of the urban sites tested.

PREVENTION AND REMEDIATION

As the old saying goes, an ounce of prevention is worth a pound of cure. This is especially true when it comes to controlling water pollution. Several important steps taken since the passage of the Clean Water Act have made surface waters today cleaner in many ways than they were 30 years ago.

For example, industrial wastes are mandated to be neutralized or broken down before being discharged to streams, lakes, and harbors. Moreover, the U.S. government has banned the production and use of certain dangerous pollutants such as DDT and PCBs.

In addition, two major changes have been introduced in the handling of sewage. First, smaller, less efficient sewage treatment plants are being replaced with modern, regional plants that include biological treatment, in which microorganisms are used to break down organic matter in the sewage. The newer plants are releasing much cleaner discharges into the receiving bodies of water (rivers, lakes, and ocean).

Second, many jurisdictions throughout the United States are building separate sewer lines for storm water and sanitary wastes. These upgrades are needed because excess water in the older, "combined" sewer systems would simply bypass the treatment process, and untreated sewage would be discharged directly into receiving bodies of water.

To minimize pollutants from nonpoint sources, the EPA is requiring all municipalities to address the problem of runoff from roads and parking lots. At the same time, the use of fertilizers and pesticides needs to be reduced. Toward this end, county extension agents are educating farmers and homeowners about their proper application and the availability of nutrient testing.

To curtail the use of expensive and potentially harmful pesticides, the approach known as *integrated pest management* can be implemented. It involves the identification of specific pest problems and the use of nontoxic chemicals and chemical-free alternatives whenever possible. For instance, aphids can be held in check by ladybug beetles and caterpillars can be controlled by applying neem oil to the leaves on which they feed.

Moreover, new urban development projects in many areas are required to implement storm-water management practices. They include such features as oil and grease traps in storm drains; swales to slow down runoff, allowing it to infiltrate back into groundwater; "wet" detention basins (essentially artificial ponds) that allow solids to settle out of runoff; and artificial wetlands that help break down contaminants in runoff. While such additions may be costly, they significantly improve water quality. They are of course much more expensive to install after those areas have been developed.

Once a waterway is polluted, cleanup is often expensive and time consuming. For instance, to increase the concentration of dissolved oxygen in a lake that has undergone eutrophication, fountains and aerators may be necessary. Specially designed boats may be needed to harvest nuisance weeds.

At times, it is costly just to identify the source of a problem. For example, if a body of water contains high levels of coliform bacteria, expensive DNA testing may be needed to determine whether the bacteria came from leakage of human sewage, pet waste, or the feces of waterfowl or other wildlife.

Contaminated sediments are sometimes difficult to treat. Available techniques range from dredging the sediments to "capping" them in place, to limit their potential exposure. Given that they act as reservoirs of pollutants, it is often best to remove the sediments and burn off the contaminants. Alternatively, the extracted sediments may be placed in confined disposal areas that prevent the pollutants from leaching back into groundwater. Dredging, however, may create additional problems by releasing pollutants back into the water column when the sediment is stirred up.

THE FUTURE OF CLEAN WATER

The EPA reports that as a result of the Clean Water Act, millions of tons of sewage and industrial waste are being treated before they are discharged into U.S. coastal waters. In addition, the majority of lakes and rivers now meet mandated water quality goals.

Yet the future of federal regulation under the Clean Water Act is unclear. In 2001, a Supreme Court decision (*Solid Waste Agency of Northern Cook County* v. *United States Army Corps of Engineers, et al.*) brought into question the power of federal agencies to regulate activities affecting water quality in smaller, nonnavigable bodies of water. This and related court decisions have set the stage for the EPA and other federal agencies to redefine which bodies of water can be protected from unregulated dumping and discharges under the Clean Water Act. As a result, individual states may soon be faced with much greater responsibility for the protection of water resources.

Worldwide, more than one billion people presently lack access to clean water sources, and over two billion live without basic sanitation facilities. A large proportion of those who die from water-related diseases are infants. We would hope that by raising awareness of these issues on an international level, the newly recognized right to clean water will become a reality for a much larger percentage of the world's population.

TRAGEDY AT MINAMATA BAY

The Chisso chemical factory, located on the Japanese island of Kyushu, is believed to have discharged between 70 and 150 tons of methylmercury (an organic form of mercury) into Minamata Bay between 1932 and 1968. The factory, a dominant presence in the region, used the chemical to manufacture acetic acid and vinyl chloride.

Methylmercury is easily absorbed upon ingestion, causing widespread damage to the central nervous system. Symptoms include numbing and unsteadiness of extremities, failure of muscular coordination, and impairment of speech, hearing, and vision. Exposure to high levels of the substance can be fatal. In addition, the effects are magnified for infants exposed to methylmercury through their mothers, both before birth and while nursing.

In the 1960s and '70s, it was revealed that thousands of Minamata Bay residents had been exposed to methylmercury. The chemical had been taken up from the bay's waters by its fish and then made its way into the birds, cats, and people who ate the fish. Consequently, methylmercury poisoning came to be called Minamata disease.

Remediation, which took as long as 14 years, involved removing the mercury-filled sediments and containing them on reclaimed land in Minamata Bay. Fish in the bay had such high levels of methylmercury that they had to be prevented from leaving the bay by a huge net, which was in place from 1974 to 1997.

Mercury poisoning has recently appeared in the Amazon basin, where deforestation has led to uncontrolled runoff of natural accumulations of mercury from the soil into rivers and streams. In the United States, testing has revealed that predator fish such as bass and walleye in certain lakes and rivers contain enough mercury to justify warnings against consuming them in large amounts.

Running Pure: The Importance of Forest Protected Areas to Drinking Water
by Nigel Dudley and Sue Stolton

THE IMPORTANCE OF FOREST PROTECTED AREAS TO DRINKING WATER
Introduction: What Do City Dwellers Need?

In the past 100 years the world population tripled, but water use for human purposes multiplied sixfold! Water is, in theory, a quintessentially renewable resource. Most of the world's surface is covered in water and over much of the world it falls, unbidden and with great regularity, from the skies. Yet, the carelessness and profligacy with which water resources have been used, the speed of human population growth and the increasing per capita demands for water together mean that provision of adequate, safe supplies of water is now a major source of concern, expense and even international tension. At the World Summit on Sustainable Development in Johannesburg in 2002, over 80 per cent of the participating decision-makers identified water as a key issue to be addressed by Heads of State from countries throughout the world.

Overall, the greatest human requirement for freshwater resources is for crop irrigation, particularly in places where farming takes place in arid regions and in the great rice paddy fields of Asia. Municipal water—the focus of the current study—accounts for less than a tenth of human water use. But the need for clean drinking water is of critical importance to the growing proportion of the world's population that live in cities. Wherever a breakdown in water supply occurs, because of disasters like earthquakes, floods, wars or civil unrest, immediate and acute problems occur and reliance on contaminated water results in the rapid spread of diseases like cholera and infant diarrhoea.

Unfortunately, for many people there is no need for a disaster to make them dependent on unclean drinking water. Today, around half of the world's population lives in towns and cities, and of this urban population one third, an estimated one billion people, live without clean water or adequate sanitation, despite these services widely being regarded as basic prerequisites of a decent life. These one billion extreme have-nots are unevenly distributed around the world. Regionally, it has been estimated that 700 million people in urban Asia, or half the urban population, do not have adequate water supplies; nor do 150 million people in Africa, again about 50 per cent of the city dwellers; with a further 120 million people, about 30 per cent of the urban population, lacking clean water in Latin America and the Caribbean. Many people die each year as a direct result. Annually, 2.2 million deaths, four per cent of all fatalities worldwide, can be attributed to inadequate supplies of clean water and sanitation.

These problems are likely to increase in the future as the current rapid processes of population growth and urbanisation continue. The average size of the world's 100 largest cities grew from around 0.2 million in 1800 to 6.2 million in 2000. In 1900, there were estimated to be just 43 cities worldwide with a population of over half a million, by 1990 this figure had risen to around 800 cities worldwide—of which some 270 had more than one million and 14 had over 10 million. These trends are likely to continue for some time. Most current estimates suggest that the world's population will grow by two billion people over the next 30 years and another billion in the following 20 years. Virtually, all of these increases will be in developing countries, the bulk of which will occur in urban areas. In India, for example, World Bank forecasts are that demand for water in the urban and industrial sectors is likely to increase by 135 percent over the next 40 years.

In many arid countries, there is already an acute supply shortage. World water withdrawals rose sixfold over the last century. It has been estimated that humanity now uses 54 per cent of accessible runoff, a figure that could rise to 70 per cent by 2005. For several countries, current reliance on nonrenewable (or only very slowly renewable) groundwater sources masks a problem that could rapidly become more acute as these are exhausted. Because of population growth, the average annual per capita availability of renewable water resources is projected to fall from 6,600 cubic metres [8,633 cubic yards] today to 4,800 cubic metres [6,278 cubic yards] in 2025. In 1998, 28 countries experienced water stress or scarcity (defined when available water is lower than 1,000 cubic meters [1,308 cubic yards] per person per year). By 2025, this number is predicted to rise to 56. As the number of people in urban areas grows, so does the demand for water, food and for irrigation in agricultural areas close to the city adding further pressures on water resources.

The demand for water, along with increasing pressures on water from pollution, urbanisation and overexploitation of aquatic resources, is also creating a biodiversity crisis in freshwaters.

Although future supply problems are expected, with a few notable exceptions the current shortfall in clean water for city dwellers is seldom to do with a real lack of supply but more related to poor distribution, inadequate treatment and to some extent also poor education and a lack of understanding about the problems. For example, up to 50 per cent of the urban water in many African cities is being wasted through leakage, theft or is otherwise unaccounted for. (Conversely Melbourne, after a seven year period of extreme drought, is still supplying its citizens with some of the best quality drinking water in the world.) Efforts are being made to address these problems. Over the past 20 years for instance more than 2.4 billion people have gained access to water supply and 600 million to sanitation. The United Nations Millennium

Summit in 2000 agreed to halve the 1.1 billion people who do not have access to safe water by 2015, as part of its Millennium Development Goals.

Cities therefore face immediate problems of access to clean water and sanitation and mounting problems of supply. In recent years, increasing interest has been taken in the opportunities for maintaining urban water supplies (and perhaps even more importantly water quality) through management of natural resources. Unfortunately, the links often come into focus when something goes wrong—most commonly when resource management upstream has downstream impacts in terms of changes in water supply, increased flooding and reduced water quality. The majority of the world's population live downstream of forested watersheds and therefore are susceptible to the costs of watershed degradation. At the same time, 28 percent of the world's forest areas are in mountains, and mountains are the source of some 60 to 80 percent of the world's freshwater resources. Hence the importance of this report.

The protected areas and protected forests identified below all play a role in providing drinking water for the world's biggest cities. More often than not this water also helps feed the people, through irrigation of crops, provide electricity through hydro-electric plants, and has a recreation, aesthetic and even religious function. Of course protected areas are just one tool in a range of watershed conservation models, that can have costs of their own, but it is hoped that by highlighting their role this report will add to the growing literature extolling the benefits of long-term protection to some of the world's most important resource areas.

Is the Freshwater in Our Country Clean?

In 2004, the United States Geological Survey (USGS), a division within the U.S. Department of the Interior, released its report on the condition of the United States' water quality. The National Water-Quality Assessment (NAWQA) Program monitored 51 streams, groundwater, and aquatic ecosystems in almost every state from 1991 to 2001. The report will be followed by another study (2002–2012) to reassess 42 of the 51 study sites. The USGS reports provide important scientific data to local, state, and government organizations to help them make environmental policy decisions.

The NAWQA study looked at the current quality of water resources, how conditions have changed over time, and major factors (such as climate, land use, and topography) that affect water quality. It found that contamination of streams and groundwater is widespread in both agricultural and urban areas. It also showed that water quality and aquatic ecosystem health are "controlled by a combination of factors, including chemical use, land use, land-management practices, and natural features, such a geology, hydrology, soils, and climate."

Also useful for planners and policymakers are the report's findings that illustrate how the interaction of surface water, groundwater, land, and air affects the flow of water and contaminants in aquatic ecosystems. Not surprisingly, water quality is most impacted in areas that have larger human populations.

The study clearly shows that water quality is still an important issue, and that the goals of the 1972 Clean Water Act, one of which was "to restore and maintain the chemical, physical, and biological integrity of the Nation's waters," have yet to be realized. The following report outlines key considerations for public officials as they make decisions that affect water resources. Proper management can dramatically improve the quality of our water and our environment.

—The Editor

Water Quality in the Nation's Streams and Aquifers

from the United States Geological Survey (USGS)

INFORMATION TO MANAGE, PROTECT, AND RESTORE WATER QUALITY

The National Water-Quality Assessment (NAWQA) Program of the U.S. Geological Survey (USGS) assesses the quality of streams, ground water, and aquatic ecosystems in major river basins and aquifer systems across the Nation. These assessments characterize the ambient water resource—the source for more than 60 percent of the Nation's drinking water and water for irrigation and industry.

During its first decade (1991–2001), NAWQA completed assessments in 51 study units, which provided baseline data and information on the occurrence of pesticides, nutrients, volatile organic compounds (VOCs), trace elements, and radon in water, and on the condition of aquatic habitats and fish, insect, and algal communities. Conditions are compared to selected benchmarks, such as for drinking-water quality and the protection of aquatic organisms. Each assessment follows a nationally consistent study design and methodology, thereby providing information about local water-quality conditions as well as providing insight on where and when water quality varies regionally and nationally. This document contains findings of regional and national interest along with examples from the study units illustrating these findings.

During its second decade (2002–2012), NAWQA plans to reassess 42 of the 51 study units. These assessments will fill critical gaps in the characterization of water-quality conditions; determine trends at many of the monitoring sites; and build upon earlier assessments that link water-quality conditions and trends to natural and human factors. . . .

The first round of NAWQA assessments indicates that while many of our Nation's waters are suitable for most uses, contaminants from nonpoint and point sources continue to affect our streams and ground water in parts of every study unit. Findings from the 51 study units show that

- contamination of streams and ground water is widespread in agricultural and urban areas, and is characterized by complex mixtures of nutrients, trace elements, pesticides, VOCs, and their chemical breakdown products, and

- water quality and aquatic-ecosystem health are controlled by a combination of factors, including chemical use, land use, land-management practices, and natural features, such as geology, hydrology, soils, and climate.

Local, state, tribal, and national stakeholders use NAWQA information to design and implement strategies for managing, protecting, and monitoring water resources in many different hydrologic and land-use settings across the Nation, such as to:

- support development of regulations, standards, and guidelines that reflect actual contaminant occurrence, including contaminant mixtures, break-down products, seasonal patterns, and variability among different settings;

- identify key sources of nonpoint pollution in agricultural and urban areas;

- prioritize geographic areas and basins in which water resources and aquatic ecosystems are most vulnerable to contamination and where improved treatment or management can have the greatest benefits;

- improve strategies and protocols for monitoring, sampling, and analysis of all hydrologic components, including the atmosphere, surface water, ground water, and biological communities;

- contribute to State assessments of beneficial uses and impaired waters (Total Maximum Daily Loads or TMDLs), strategies for source-water protection and management, pesticide and nutrient management plans, and fish-consumption advisories; and,

- sustain the health of aquatic ecosystems through improved stream protection and restoration management.

NAWQA studies indicate that contaminants are widespread, albeit often at low concentrations, in river basins and aquifer systems across a wide range of landscapes and land uses. In the mostly agricultural Lower Tennessee River Basin, for example, 52 different pesticides were detected in streams and rivers, and VOCs were detected in about 67 percent of sampled springs and wells that tap underlying carbonate aquifers. Nationally, at least one pesticide was found in about 94 percent of water samples and in 90 percent of fish samples from streams, and in about 55 percent of shallow wells sampled in agricultural and urban areas.

The type and concentrations of contaminants that are found in urban and agricultural water resources are closely related to the chemicals that are used (such as fertilizers and pesticides) or that are released with waste products (such as sewage or manure). For example, phosphorus and many insecticides, such as diazinon, carbaryl, chlorpyrifos, and malathion, were detected more frequently and usually at higher concentrations in urban streams than in agricultural streams. Nationally, at least one pesticide guideline established to protect aquatic life was exceeded in nearly all (about 93 percent) of the urban streams sampled.

Nitrogen and many herbicides—most commonly atrazine and its breakdown product deethylatrazine (DEA), metolachlor, alachlor, and cyanazine—generally were detected more frequently and usually at higher concentrations in streams and shallow ground water in agricultural areas than in urban areas. Occurrence is linked to use; these herbicides rank in the top five used for agriculture. Concentrations of nitrate exceeded the U.S. Environmental Protection Agency (USEPA) drinking-water standard of 10 milligrams per liter in samples collected from about 20 percent of shallow wells in agricultural areas (versus about 3 percent in urban areas).

VOCs were detected frequently in shallow ground water beneath urban areas (in about 90 percent of monitoring wells sampled) and less frequently in shallow ground water beneath agricultural areas (in 20 percent of monitoring wells). Some of the most common VOCs in urban areas, such as in the Delaware River Basin, were the solvents trichloroethene (TCE), tetrachloroethene (PCE), 1,1,1-trichloroethane (TCA), and trichloromethane (also known as chloroform), which is also a disinfection by-product of water treatment; and the gasoline-related compounds benzene, toluene, xylene, and methyl tert-butyl ether (MTBE).

New pesticides, VOCs, and other synthetic chemicals are introduced into the environment every year as products are approved for agricultural or urban use. USGS has expanded its laboratory methods to analyze for these "emerging" contaminants, testing new methods in areas where their use is the greatest.

WATER QUALITY AND ITS CONNECTION TO LAND USE— AGRICULTURAL AND URBAN SOURCES OF NONPOINT POLLUTION
Implications
Reducing chemical use and improving disposal practices can help reduce contaminant concentrations in both urban and agricultural settings. More information about chemical use— data that are virtually unavailable in urban areas—is critical to linking contaminants to their sources, thus helping individuals, businesses, and industry as well as local, State, and Federal governments to improve water quality.

INTERACTIONS AMONG WATER, AIR, AND AQUATIC ECOSYSTEMS
Interactions among surface water, ground water, land, and the atmosphere govern the occurrence and movement of water

and contaminants, and thus affect the health of aquatic ecosystems. NAWQA studies therefore include assessments of multiple components of the hydrologic system, including streams, rivers, and ground water within the study units; the atmosphere; and biological communities (primarily fish, aquatic invertebrates, and algae) and stream habitat.

INTERACTIONS BETWEEN SURFACE WATER AND GROUND WATER

Ground-water discharge can influence the quality of streams and ultimately the receiving water. For example, ground water supplies about half of the water and nitrogen to streams in the Chesapeake Bay watershed and is therefore an important source of nitrogen to the Chesapeake Bay. Streams, in turn, can influence the quality of ground water. Contaminants generally are more prevalent and detected at higher concentrations in streams than in ground water. This is largely determined by chemical properties of contaminants and flow conditions— ground water typically is not vulnerable to contamination by compounds that attach to soils or that are unstable in water, and relatively long residence times along ground-water flow paths allow many chemicals to degrade, disperse, or be diluted before reaching a well.

Interactions between surface water and ground water are affected by natural features such as soils, geology, and hydrology, and by human activities such as ground-water pumping. Exchanges of water and contaminants can be rapid in areas underlain by carbonate rocks or by permeable and well-drained soils and sediment, such as sand and gravel. Pumping ground water can accelerate exchanges between streams and ground water. In some areas, concentrations of contaminants can decrease during these exchanges, such as when chemicals are sorbed onto soil particles or are transformed by chemical or biological processes.

Ground-water and surface-water exchanges are substantial in carbonate aquifers near San Antonio, Texas. Major streams lose water to the Edwards aquifer as they flow across the highly permeable, faulted and fractured carbonate rocks of the aquifer outcrop. Although streams that recharge the aquifer originate in and flow through mostly undeveloped range-land, some of them also flow through urbanized northern San Antonio.

Implications

Ground-water contributions to streams and rivers can be substantial. For this reason, ground-water flow and quality should be considered in making decisions for stream protection, such as in establishing Total Maximum Daily Loads (TMDLs). Surface-water recharge to ground water also is important, particularly where public-supply wells are located near streams, and should be considered in source-water management and well-head protection programs.

RIVERS, STREAMS, AND WETLANDS

Why Are Small Streams and Wetlands Important?

The small streams and wetlands that are the sources of rivers have ecological functions that are distinct from larger systems. Because water flows downhill, what happens where a river begins affects the quality of water farther down the river. The science of this was recognized in the Clean Water Act of 1972, which included the small streams and wetlands in its definition of waters to be protected. Unfortunately, the fate of many of these streams and wetlands is unclear as different government administrations change the definitions of which wetlands and waterways should be protected and which can be filled. And as more streams and wetlands are filled or polluted, those that remain are forced to play an even larger role.

Why, then, do small streams and wetlands matter? Some small streams and wetlands act as filters for pollution and sediment and recharge groundwater. Headwater streams and small wetlands provide habitats for specialized plant and animal communities.

According to the following selection from *Where Rivers Are Born: The Scientific Imperative for Defending Small Streams and Wetlands*, a report written by 11 experts from the American Rivers and Sierra Club, small headwater streams represent about three-quarters of the stream and river channels in the United States. A problem is that many of these streams do not appear on United States Geological Survey (USGS) maps, as the state of Ohio found. In doing its own survey, the state found that headwater streams (smaller streams that feed into rivers) make up 80% of Ohio's total length of streams, yet did not show up on any USGS maps. The following report discusses the growing volume of scientific data that show just how valuable small streams and the wetlands around them really are.

—The Editor

Where Rivers Are Born:
The Scientific Imperative for
Defending Small Streams and Wetlands
from the American Rivers and Sierra Club

Our nation's network of rivers, lakes and streams originates from a myriad of small streams and wetlands, many so small they do not appear on any map. Yet these headwater streams and wetlands exert critical influences on the character and quality of downstream waters. The natural processes that occur in such headwater systems benefit humans by mitigating flooding, maintaining water quality and quantity, recycling nutrients, and providing habitat for plants and animals. This paper summarizes the scientific basis for understanding that the health and productivity of rivers and lakes depends upon intact small streams and wetlands.

Historically, federal agencies have interpreted the protections of the Clean Water Act to cover all the waters of the United States, including small streams and wetlands. Despite this, many of these ecosystems have been destroyed by agriculture, mining, development and other human activities. The extent to which small streams and wetlands should remain under the protection of the Clean Water Act is currently (2003) under consideration in federal agencies and Congress. Extensive scientific studies document the significance of these small systems and form the basis for this paper. Further references are provided at the end of the document.

We know from local/regional studies that small, or headwater, streams make up at least 80 percent of the nation's stream network. However, scientists' abilities to extend these local and regional studies to provide a national perspective is hindered by the absence of a comprehensive database that catalogs the full extent of streams in the United States. The topographic maps most commonly used to trace stream networks do not show most of the nation's headwater streams and wetlands. Thus, such maps do not provide detailed enough information to serve as a basis for stream protection and management. Scientists often refer to the benefits humans receive from the natural functioning of ecosystems as ecosystem services. The special physical and biological characteristics of intact small streams and wetlands provide natural flood control, recharge groundwater, trap sediments and pollution from fertilizers, recycle nutrients, create and maintain biological diversity, and sustain the biological productivity of downstream rivers, lakes and estuaries. These ecosystem services are provided by seasonal as well as perennial streams and wetlands. Even when such systems have no visible overland connections to the stream network, small streams and wetlands are usually linked to the larger network through groundwater. Small streams and wetlands offer an enormous array of habitats for plant, animal and microbial life. Such small freshwater systems provide shelter, food, protection from predators, spawning sites and nursery areas, and travel corridors through the landscape. Many species depend on small streams and wetlands at some point in their life history. For example, headwater streams are vital for maintaining many of America's fish species, including trout and salmon. Both perennial and seasonal streams and wetlands provide valuable habitat. Headwater streams and wetlands also provide a rich resource base that contributes to the productivity of both local food webs and those farther downstream. However, the unique and diverse biota of headwater systems is increasingly imperiled. Human-induced changes to such waters, including filling streams and wetlands, water pollution, and the introduction of exotic species, can diminish the biological diversity of such small freshwater systems, thereby also affecting downstream rivers and streams.

Because small streams and wetlands are the source of the nation's fresh waters, changes that degrade these headwater systems affect streams, lakes, and rivers downstream. Land-use changes in the vicinity of small streams and wetlands can impair the natural functions of such headwater systems. Changes in surrounding vegetation, development that paves and hardens soil surfaces, and the total elimination of some small streams reduces the amount of rainwater, runoff and snowmelt the stream network can absorb before flooding. The increased volume of water in small streams scours stream channels, changing them in a way that promotes further flooding. Such altered channels have bigger and more frequent floods. The altered channels are also less effective at recharging groundwater, trapping sediment, and recycling nutrients. As a result, downstream lakes and rivers have poorer water quality, less reliable water flows, and less diverse aquatic life. Algal blooms and fish kills can become more common, causing problems for commercial and sport fisheries. Recreational uses may be compromised. In addition, the excess sediment can be costly, requiring additional dredging to clear navigational channels and harbors and increasing water filtration costs for municipalities and industry.

The natural processes that occur in small streams and wetlands provide Americans with a host of benefits, including flood control, adequate highquality water, and habitat for a variety of plants and animals. Scientific research shows that healthy headwater systems are critical to the healthy functioning of downstream streams, rivers, lakes and estuaries. To provide the ecosystem services that sustain the health of our nation's waters, the hydrological, geological, and biological characteristics of small streams and wetlands require protection. . . .

SMALL STREAMS AND WETLANDS PROVIDE BENEFICIAL ECOSYSTEM SERVICES

Natural processes that occur in small streams and wetlands provide humans with a host of benefits, including flood control, maintenance of water quantity and quality, and habitat for a variety of plants and animals. For headwater streams and wetlands to provide ecosystem services that sustain the health of our nation's waters, the hydrological, geological and biological components of stream networks must be intact.

Small Streams and Wetlands Provide Natural Flood Control

Floods are a natural part of every river. In times past, waters of the Mississippi River routinely overtopped its banks.

Floodwaters carried the sediment and nutrients that made the Mississippi Delta's soil particularly suitable for agriculture. But floods can also destroy farms, houses, roads and bridges.

When small streams and wetlands are in their natural state, they absorb significant amounts of rainwater, runoff and snowmelt before flooding. However, when a landscape is altered, such as by a landslide or large forest fire or a housing development, the runoff can exceed the absorption capacity of small streams. Moreover, the power of additional water coursing through a channel can change the channel itself. Humans often alter both landscape and stream channels in ways that result in larger and more frequent floods downstream.

A key feature of streams and rivers is their shape. Unlike a concrete drainage ditch, a natural streambed does not present a smooth surface for water flow. Natural streambeds are rough and bumpy in ways that slow the passage of water. Particularly in small narrow streams, friction produced by a stream's gravel bed, rocks, and dams of leaf litter and twigs slows water as it moves downstream. Slower moving water is more likely to seep into a stream's natural water storage system—its bed and banks—and to recharge groundwater. Slower moving water also has less power to erode stream banks and carry sediment and debris downstream.

In watersheds that are not carefully protected against impacts of land development, stream channels often become enlarged and incised from increased runoff. Changed channels send water downstream more quickly, resulting in more flooding. For example, after forests and prairies in Wisconsin watersheds were converted to agricultural fields, the size of floods increased. This change in land use had altered two parts of the river systems' equation: the amount of runoff and shape of the stream channel. Cultivation destroyed the soil's natural air spaces that came from worm burrows and plant roots. The resulting collapse of the soil caused more rainfall to run off into streams instead of soaking into the ground. Additional surface runoff then altered the stream channels, thereby increasing their capacity to carry large volumes of water quickly downstream. These larger volumes flow downstream at much higher velocity, rather than soaking into the streambed.

Urbanization has similar effects; paving previously-vegetated areas leads to greater storm runoff, which changes urban stream channels and ultimately sends water more quickly downstream. Covering the land with impermeable surfaces, such as roofs, roads, and parking lots, can increase by several times the amount of runoff from a rainstorm. If land uses change near headwater streams, effects are felt throughout the stream network. In an urban setting, runoff is channeled into storm sewers, which then rapidly discharge large volumes of water into nearby streams. The additional water causes the stream to pick up speed, because deeper water has less friction with the streambed. The faster the water moves, the less it can soak into the streambed and banks. Faster water also erodes channel banks and beds, changing the shape of a channel. The effect is magnified downstream, because larger rivers receive water from tens, sometimes hundreds, of small headwater basins. When such changes are made near headwater streams, downstream portions of the stream network experience bigger and more frequent flooding.

As regions become more urbanized, humans intentionally alter many natural stream channels by replacing them with storm sewers and other artificial conduits. When larger, smoother conduits are substituted for narrow, rough-bottomed natural stream channels, flood frequency increases downstream. For example, three decades of growth in storm sewers and paved surfaces around Watts Branch Creek, Maryland more than tripled the number of floods and increased average annual flood size by 23 percent.

Small Streams and Wetlands Maintain Water Supplies

Headwater systems play a crucial role in ensuring a continual flow of water to downstream freshwater ecosystems. Water in streams and rivers comes from several sources: water held in the soil, runoff from precipitation, and groundwater. Water moves between the soil, streams and groundwater. Wetlands, even those without any obvious surface connection to streams, are also involved in such exchanges by storing and slowly releasing water into streams and groundwater, where it later resurfaces at springs. Because of these interactions, groundwater can contribute a significant portion of surface flow in streams and rivers; conversely, surface waters can also recharge groundwater. If connections between soil, water, surface waters, and groundwater are disrupted, streams, rivers, and wells can run dry. Two-thirds of Americans obtain their drinking water from a water system that uses surface water. The remaining one-third of the population relies on groundwater sources. The quality and amount of water in both of these sources respond to changes in headwater streams.

USGS estimates that, on average, from 40 to 50 percent of water in streams and larger rivers comes from groundwater. In drier regions or during dry seasons, as much as 95 percent of a stream's flow may come from groundwater. Thus, the recharge process that occurs in unaltered headwater streams and wetlands both moderates downstream flooding in times of high water and maintains stream flow during dry seasons.

Headwater streams and wetlands have a particularly important role to play in recharge. These smallest upstream components of a river network have the largest surface area of soil in contact with available water, thereby providing the greatest opportunity for recharge of groundwater. Moreover, water level in headwater streams is often higher than the water table, allowing water to flow through the channel bed and banks into soil and groundwater. Such situations occur when water levels are high, such as during spring snowmelt or rainy seasons. During dry times, the situation in some reaches of the stream network, particularly those downstream, may reverse, with water flowing from the soil and groundwater through the channel banks and bed into the stream. This exchange of water from the soil and groundwater into the stream maintains stream flow. However, if land-use changes increase the amount of precipitation that runs off into a stream rather than soaking into the ground, the recharge process gets short-circuited. This increased volume of stream water flows rapidly downstream rather than infiltrating into soil and groundwater. The consequence is less overall groundwater recharge, which often results in less water in streams during drier seasons.

Therefore, alteration of small streams and wetlands disrupts the quantity and availability of water in a stream and river system. Protecting headwater streams and wetlands is important for maintaining water levels needed to support everything from fish to recreational boating to commercial ship traffic.

Small Streams and Wetlands Trap Excess Sediment

Headwater systems retain sediment. Like the flow of water, movement of sediment occurs throughout a river network. Thus, how a watershed is managed and what kinds of land uses occur there have substantial impact on the amount of sediment delivered to larger rivers downstream. Increased sediment raises water purification costs for municipal and industrial users, requires extensive dredging to maintain navigational channels, and degrades aquatic habitats. Intact headwater streams and wetlands can modulate the amount of sediment transported to downstream ecosystems.

Runoff from rain, snowmelt and receding floodwaters can wash soil, leaves and twigs into streams, where the various materials get broken up into smaller particles or settle out. If natural vegetation and soil cover are disturbed by events and activities such as fires, farming or construction, runoff increases, washing more materials into streams. At the same time, the increased velocity and volume of water in a stream cause erosion within the streambed and banks themselves, contributing additional sediment to the stream system. Moreover, the faster, fuller stream can carry more and larger chunks of sediment further downstream.

One study found that land disturbances such as urban construction can, at minimum, double the amount of sediment entering headwater streams from a watershed. A Pennsylvania study showed how, as a 160-acre [65-hectare] headwater watershed became more urbanized, channel erosion of a quartermile

[0.4-km] stretch of stream generated 50,000 additional cubic feet [1,416 cubic meters] of sediment in one year—enough to fill 25 moderate-sized living rooms. In a nonurban watershed of the same size, it would take five years to generate the same amount of sediment. Such studies demonstrate that landscape changes such as urbanization or agriculture, particularly without careful protection of headwater streams and their riparian zones, may cause many times more sediment to travel downstream.

Excess Sediment in Downstream Ecosystems Costs Money

Keeping excess sediment out of downstream rivers and lakes is one ecosystem service intact small streams and wetlands provide. Once sediment moves further downstream, it becomes an expensive problem. Too much sediment can fill up reservoirs and navigation channels, damage commercial and sport fisheries, eliminate recreation spots, harm aquatic habitats and their associated plants and animals, and increase water filtration costs.

Additional sediment damages aquatic ecosystems. Sediment suspended in the water makes it murkier; as a result, underwater plants no longer receive enough light to grow. Fish that depend on visual signals to mate may be less likely to spawn in murky water, thereby reducing fish populations. High levels of sediment suspended in water can even cause fish kills. Even as it settles to the bottom, sediment continues to cause problems because it fills the holes between gravel and stones that some animals call home, smothers small organisms that form the basis of many food webs, and can also smother fish eggs.

Getting rid of sediment is expensive. For example, keeping Baltimore Harbor navigable costs $10 to $11.5 million annually to dredge and dispose of sediment the Patapsco River deposits in the harbor.

Small Streams and Wetlands Retain Sediment

Headwater streams and wetlands typically trap and retain much of the sediment that washes into them. The faster the water travels, the larger the particles it can carry. So, natural obstructions in small streams, rocks, downed logs, or even just a bumpy stream bottom slow water and cause sediment to settle out of the water column. Wetlands, whether or not they have a surface connection to a nearby stream, are often areas where runoff slows and stops, dropping any debris the water may be carrying. Because headwater streams represent 75 percent or more of total stream length in a stream network, such streams and their associated wetlands retain a substantial amount of sediment, preventing it from flowing into larger rivers downstream.

Even ephemeral streams can retain significant amounts of sediment. Such small headwater streams expand and contract in response to heavy rains. During expansion, a stream flows over what was a dry or damp streambed. Most of the water at the leading edge of a growing stream, called the "trickle front," soaks into the streambed and does not carry sediment downstream. In a small watershed near Corvallis, Oregon, researchers found that 60 to 80 percent of sediment generated from forest roads traveled less than 250 feet [76 meters] downstream before settling out in stream pools. Headwater streams can store sediment for long periods of time: research in Oregon's Rock Creek basin found that headwater streams could retain sediment for 114 years.

Natural Cleansing Ability of Small Streams and Wetlands Protects Water Quality

Materials that wash into streams include everything from soil, leaves and dead insects to runoff from agricultural fields and animal pastures. One of the key ecosystem services that stream networks provide is the filtering and processing of such materials. Healthy aquatic ecosystems can transform natural materials like animal dung and chemicals such as fertilizers into less harmful substances. Small streams and their associated wetlands play a key role in both storing and modifying potential pollutants, ranging from chemical fertilizers to rotting salmon carcasses, in ways that maintain downstream water quality.

Excess Nutrients Cause Problems in Rivers and Lakes

Inorganic nitrogen and phosphorus, the main chemicals in agricultural fertilizers, are essential nutrients not just for plants, but for all living organisms. However, in excess or in the wrong proportions, these chemicals can harm natural systems and humans.

In freshwater ecosystems, eutrophication, the enriching of waters by excess nitrogen and phosphorus, reduces water quality in streams, lakes, estuaries and other downstream water bodies. One obvious result is the excessive growth of algae. More algae clouds previously clear streams, such as those favored by trout. In addition to reducing visibility, algal blooms reduce the amount of oxygen dissolved in the water, sometimes to a degree that causes fish kills. Fish are not the only organisms harmed: some of the algae species that grow in eutrophic waters generate tastes and odors or are toxic, a clear problem for stream systems that supply drinking water for municipalities. In addition, increased nitrogen can injure people and animals. Excess nitrogen in the form called nitrate in drinking water has been linked to "blue baby disease" (methemoglobinemia) in infants and also has toxic effects on livestock.

HEADWATER STREAMS TRANSFORM AND STORE EXCESS NUTRIENTS

Headwater streams and associated wetlands both retain and transform excess nutrients, thereby preventing them from traveling downstream. Physical, chemical and biological processes in headwater streams interact to provide this ecosystem service.

Compared with larger streams and rivers, small streams, especially shallow ones, have more water in physical contact with a stream channel. Therefore, the average distance traveled by a particle before it is removed from the water column is shorter in headwater streams than in larger ones. A study of headwater streams in the southern Appalachian Mountains found that both phosphorus and the nitrogen-containing compound ammonium traveled less than 65 feet [20 meters] downstream before being removed from the water.

In headwater streams and wetlands, more water is in direct contact with the streambed, where most processing takes place. Bacteria, fungi and other microorganisms living on the bottom of a stream consume inorganic nitrogen and phosphorus and convert them into less harmful, more biologically beneficial compounds. A mathematical model based on research in 14 headwater streams throughout the U.S. shows that 64 percent of inorganic nitrogen entering a small stream is retained or transformed within 1,000 yards [914 meters].

Channel shape also plays a role in transforming excess nutrients. Studies in Pennsylvania have shown that when the forest surrounding headwaters is replaced by meadows or lawns, increased sunlight promotes growth of grasses along stream banks. The grasses trap sediments, create sod, and narrow the stream channel to one-third of the original width. Such narrowing reduces the amount of streambed available for microorganisms that process nutrients. As a result, nitrogen and phosphorus travel downstream five to ten times farther, increasing risks of eutrophication.

Streams do not have to flow yearround to make significant contributions to water quality. Fertilizers and other pollutants enter stream systems during storms and other times of high runoff, the same times that ephemeral and intermittent streams are most likely to have water and process nutrients. Federal, state and local programs spend considerable sums of money to reduce non-point source inputs of nutrients because they are a major threat to water quality. One principal federal program, the EPA's 319 cost-share program, awarded more than $1.3 billion between 1990 and 2001 to states and territories for projects to control non-point pollution. Failure to maintain nutrient removal capacity of ephemeral and intermittent streams and wetlands would undermine these efforts.

Wetlands also remove nutrients from surface waters. Several studies of riparian wetlands have found that those associated with the smallest streams to be most effective in removing nutrients from surface waters. For example, headwater wetlands comprise 45 percent of all wetlands able to improve water quality in four Vermont watersheds. Another study found that wetlands associated with first-order streams are responsible for 90 percent of wetland phosphorus removal in eight northeastern watersheds. Such studies demonstrate that riparian wetlands, especially those associated with small streams, protect water quality.

As land is developed, headwater streams are often filled or channeled into pipes or paved waterways, resulting in fewer and shorter streams. For example, as the Rock Creek watershed in Maryland was urbanized, more than half of the stream channel network was eliminated. In even more dramatic fashion, mining operations in the mountains of central Appalachia have removed mountain tops and filled valleys, wiping out entire headwater stream networks. From 1986 to 1998, more than 900 miles [1,448 km] of streams in central Appalachia were buried, more than half of them in West Virginia.

If headwater streams and wetlands are degraded or filled, more fertilizer applied to farm fields or lawns reaches larger downstream rivers. These larger rivers process excess nutrients from fertilizer much more slowly than smaller streams. Losing the nutrient retention capacity of headwater streams would cause downstream water bodies to contain higher concentrations of nitrogen and phosphorus. A likely consequence of additional nutrients would be the further contamination and eutrophication of downstream rivers, lakes, estuaries and such waters as the Gulf of Mexico.

Natural Recycling in Headwater Systems
Sustains Downstream Ecosystems

Recycling organic carbon contained in the bodies of dead plants and animals is a crucial ecosystem service. Ecological processes that transform inorganic carbon into organic carbon and recycle organic carbon are the basis for every food web on the planet. In freshwater ecosystems, much of the recycling happens in small streams and wetlands, where microorganisms transform everything from leaf litter and downed logs to dead salamanders into food for other organisms in the aquatic food web, including mayflies, frogs and salmon.

Like nitrogen and phosphorus, carbon is essential to life but can be harmful to freshwater ecosystems if it is present in excess or in the wrong chemical form. If all organic material received by headwater streams and wetlands went directly downstream, the glut of decomposing material could deplete oxygen in downstream rivers, thereby damaging and even killing fish and other aquatic life. The ability of headwater streams to transform organic matter into more usable forms helps maintain healthy downstream ecosystems.

HEADWATER STREAM SYSTEMS STORE AND TRANSFORM EXCESS ORGANIC MATTER

Intact headwater systems both store and process organic matter in ways that modulate the release of carbon to downstream lakes and rivers. Headwater systems receive large amounts of organic matter, which can be retained and transformed into more palatable forms through decomposition processes. This organic matter is anything of biological origin that falls into, washes into or dies in a stream. Plant parts, such as leaves, twigs, stems and larger bits of woody debris, are the most common of these items. Another source of organic material is dead stream organisms, such as bits of dead algae and bacteria or bodies of insects and even larger animals. Waste products of plants and animals also add organic carbon to water. Water leaches dissolved organic carbon from organic materials in a stream and watershed like tea from a tea bag.

Much of the organic matter that enters headwater systems remains there instead of continuing downstream. One reason is that the material often enters headwater streams as large pieces, such as leaves and woody debris, that are not easily carried downstream. In addition, debris dams that accumulate in headwater streams block the passage of materials. One study found four times more organic matter on the bottoms of headwater streams in forested watersheds than on the bottoms of larger streams. Another reason material stays in headwater streams is that food webs in small streams and wetlands process organic matter efficiently. Several studies have found that headwater streams are far more efficient at transforming organic matter than larger streams.

For example, one study showed that, for a given length of stream, a headwater stream had an eight-fold higher processing efficiency than a fourth-order channel downstream. Microorganisms in headwater stream systems use material such as leaf litter and other decomposing material for food and, in turn, become food for other organisms. For example, fungi that grow on leaf litter become nutritious food for invertebrates that make their homes on the bottom of a stream, including mayflies, stoneflies and caddis flies. These animals provide food for larger animals, including birds such as flycatchers and fish such as trout.

Headwater Systems Supply Food for Downstream Ecosystems

The organic carbon released by headwater streams provides key food resources for downstream ecosystems. Headwater ecosystems control the form, quality and timing of carbon supply downstream. Although organic matter often enters headwaters in large amounts, such as when leaves fall in autumn or storm runoff carries debris into the stream, those leaves and debris are processed more slowly. As a result, carbon is supplied to downstream food webs more evenly over a longer period of time. Forms of carbon delivered range from dissolved organic carbon that feeds microorganisms to the drifting insects such as mayflies and midges that make ideal fish food. Such insects are the preferred food of fish such as trout, char and salmon. One study estimated that fishless headwater streams in Alaska export enough drifting insects and other invertebrates to support approximately half of the fish production in downstream waters.

Processed organic matter from headwater streams fuels aquatic food webs from the smallest streams to the ocean. Only about half of all first-order streams drain into second-order streams; the other half feed directly into larger streams or directly into estuaries and oceans, thus delivering their carbon directly to these larger ecosystems. The health and productivity of downstream ecosystems depends on processed organic carbon—ranging from dissolved organic carbon to particles of fungus, and leaf litter to mayflies and stoneflies—delivered by upstream headwater systems.

Why Does Phosphorus Cause Water Pollution Around the World?

If you look at a container of plant fertilizer, you will see a "P," which stands for "phosphorus," an essential nutrient in plant growth. Too much phosphorus in rivers and lakes causes algae to bloom. Then, the lake or river is said to be "eutrophic." The condition is easy to identify. The water becomes covered with a slimy green blanket of algae. This increased plant growth increases the biological oxygen demand in the water, and the system can "choke" from a lack of oxygen. Dead fish then float on the water. Eutrophication can cost millions of dollars through lost fisheries and recreational opportunities. In Wisconsin, the eutrophication of Lake Mendota has been estimated to cost $50 million in lost recreation and devalued properties.

Phosphorus is a mineral that is found in rocks and soil. It is cycled continuously around the planet. How, then, does a river or lake get more phosphorus than it needs? Phosphorus is mined and then concentrated in fertilizers and animal feeds. Through rains and irrigation, the phosphorus runs off agricultural fields or lawns into storm drains and into rivers or lakes. This is called nonpoint source pollution. Much has been done to control the point sources of phosphorus from detergents and sewage. But nonpoint sources are more difficult to control.

Water pollution is a problem around the world, as outlined in the following article, "P Soup: The Global Phosphorus Cycle." The most recent data (1994) shows that eutrophication is a problem in about 50% of lakes and reservoirs in Southeast Asia, Europe, and North America. In the Gulf of Mexico, there exists an extensive "dead zone" in which nutrient run-off from agriculture has caused large numbers of ocean species to die out.

There are steps that can be taken to protect the water quality. As more scientific studies show the connection between land and water, planners are seeing the economic advantage of protecting land in order to protect water. New York City officials purchased watershed land to naturally filter the city's water supply instead of building a water filtration plant. The city saved millions of dollars, while also protecting plant and animal habitats. Thus, as with any environmental solution, solving one problem also helps solve others.

—The Editor

P Soup: The Global Phosphorus Cycle
by Elena Bennett and Steve Carpenter

Think of global environmental change, and you'll probably think most immediately of such sweeping atmospheric phenomena as global warming or ozone depletion. Many of the other environmental disruptions we're familiar with—toxic dumps, decimated forests, eroded fields—seem largely confined to particular localities.

Yet there are some environmental changes that, while appearing to be locally confined, are in fact manifestations of worrisome global patterns. Look at the algae forming on a local farm pond, for example, and you're seeing the result of a process—the phosphorus cycle—that extends far beyond that farm.

Algae thrives (literally "blooms") on runoff of waste fertilizer or other materials containing phosphorus. While human-caused changes in the closely related nitrogen cycle have been widely publicized, impacts on the phosphorus cycle are less well known. Our research suggests, however, that the movement of phosphorus is indeed a global phenomenon—and that that patch of algae you see in the pond at your feet may be affected by changes in the soil hundreds or thousands of miles away.

Both nitrogen and phosphorus are essential nutrients for plants and are therefore present in most fertilizers in addition to being present in agricultural and municipal waste products. As a result, the movement of large amounts of fertilizers around the planet can also mean the movement of excessive nutrients from one place to another. Typically, some of the fertilizer used on a farm does not stay there but moves downhill where it can get into a downstream aquatic ecosystem—a river, lake, or bay. Concentrations of excess nutrients in these bodies of water cause the patches of algae to expand prolifically. Such "eutrophication," as the green blanketing of the water is called, can be a crippling process: it suffocates the life under the slime—killing fish, diminishing biodiversity, and emitting noxious odors. It reduces the value of the water for most human uses—whether for drinking, fishing, swimming, or even boating.

Lake Mendota, in Madison, Wisconsin is a classic example of a eutrophic lake in an urbanizing, but primarily agricultural setting. The lake has exhibited many of the symptoms associated with eutrophication since agriculture became the primary land cover in the surrounding watershed, in place of the native prairie and oak savannah. Blooms of blue-green algae have been common here since the 1880s. Along with these blooms came dramatic changes in the food web, including loss of some native species and increased populations of non-native species such as Eurasian milfoil and carp. Eutrophication has greatly diminished the lake's recreational value.

Historically, Madison has never developed public swimming pools, because people were always able to swim in the area's plentiful lakes. Now, as eutrophication worsens, there is increasing public pressure to develop pools because many of the lakes are no longer swimmable. For Lake Mendota alone, the cost of eutrophication has been estimated to be about $50 million in lost recreation and property values. Even so, what happened to this lake is seen as a local story—of little interest to someone in North Carolina or South China. If the people in Madison want to deal with this problem, obviously it is their own local farms and sewer pipes they have to deal with. What's not so obvious is that what has happened here results from massive changes in the flow of phosphorus around the globe.

HUMAN IMPACT ON THE GLOBAL PHOSPHORUS CYCLE

Long before humans arrived on the scene, phosphorus was moving around the planet in a natural cycle that probably took millions of years to complete. A phosphorus molecule might be trapped in rock, then released by erosion to start its gravity-driven journey to the ocean. Along the way, it might be taken up by plants and then animals, then returned to the soil or water via dead vegetation or urine, to continue its slow trek downhill until it finally reached the ocean. Once in the ocean—probably after taking more detours through plants and animals along the way—it would sink into the sediment. In time, geological processes would turn this sediment to terrestrial rock—reincorporating the phosphorus molecule. The cycle would begin again.

With the advent of human agriculture and urbanization, the natural cycle was in some respects shortcircuited. Technological advances, especially in the past 50 years, enabled us to mine phosphorus on a large scale, make fertilizer and other products from it, and transport these products around the world, dramatically accelerating the now not-so-natural phosphorus cycle. Globally, we estimate that the annual accumulation of phosphorus in the Earth's freshwater and terrestrial ecosystems has almost quadrupled, from around 3.5 terragrams per year before humans began mining and farming on a large scale, to around 13 terragrams per year now.

To understand the cycle as it was before human interventions began, it is useful to think of the cycle as a flow from the earth's crust back to earth's crust, through four main compartments. First, phosphorus-containing rock is weathered—worn by wind, rain, freezing and thawing, etc.—until it becomes soil. With the soil, it moves into lakes and rivers, which transport it to the ocean. Some of it is dissolved in water, and some adsorbed to soil particles that are carried by erosion down to the sea and, ultimately, to the ocean bottom. There it awaits the tectonic movement that will lift up the rock and make it part of the land again. The cycle is described by Aldo Leopold in his essay, "Odyssey," in *A Sand County Almanac*: "X had marked time in the limestone ledge since the Paleozoic seas covered the land. Time, to an atom locked in a rock, does not pass. The break came when a bur-oak root nosed down a crack and began prying and sucking. In the flash of a century the rock decayed, and X was pulled out and up into the world of living things. He helped build a flower, which became an acorn, which fattened a deer, which fed an Indian, all in a single year. . . ." The narrative continues through a journey of many adventures involving a bluestem, a plover, some phlox, a fox, a buffalo, a spiderwort, a prairie fire, another fox, another Indian, a beaver, a bayou, and a riverbank. Then, "One spring an oxbow caved the bank and after one short week of freshet [a rise or overflow of a stream] X lay again in his ancient prison, the sea."

Within each of the four compartments of this cycle—the Earth's crust, the soil, aquatic systems and the oceans—there are more rapid cycles of phosphorus through the biosphere. In the soil phase, the phosphorus does not all just stay put in the ground.

A Brief History of Phosphorus

In his book *The 13th Element: The Sordid Tale of Murder, Fire, and Phosphorus*, science writer John Emsley tells the story of a material that acquired a notorious reputation over three centuries—in making the nerve gas ethyl S-2 diisopropylaminoethyl methylphosphrothiolate (VX); in the organophosphate insecticides Tetraethyl diphosphate, Parathion, and Malathion; in mortar and howitzer shells; in the bombs Hitler rained on Britain; in the execution of numerous murders; and—allegedly—in the infliction of Gulf War Syndrome. But from the start, as suggested by Emsley's account of its discovery, the users of phosphorus had little inkling of the damage its disruption might ultimately do:

> Uncertainty still surrounds the date on which phosphorus was first made. We can be fairly sure the place was Hamburg in Germany, and that the year was probably 1669, but the month and day are not recorded, though it must have been night time. The alchemist who made the discovery stumbled upon a material the like of which had never been seen. Unwittingly he unleashed upon an unsuspecting world one of the most dangerous materials ever to have been made.

Some of it is taken up from the soil by plants, which are eaten by animals, which may in turn be eaten by other animals, before the phosphorus is eventually returned to the soil in manure or through decomposition of the animals' bodies after they die. Of course, the phosphorus molecule can go through several more of these rapid biospheric cycles before moving to the next compartment of the global cycle. Similar biospheric cycles may then take place in the ocean.

People do several things that impact both the larger global cycle and the smaller, more rapid cycles of phosphorus through the biosphere. Most phosphorus mining takes place in only a few locations around the world—primarily in Florida, West and North Africa, and Russia. Mined phosphorus is then made into fertilizers, animal feeds, and other

On that dark night our lone alchemist was having no luck with his latest experiments to find the philosopher's stone. Like many before him he had been investigating the golden stream, urine, and he was heating the residues from this which he had boiled down to a dry solid. He stoked his small furnace with more charcoal and pumped the bellows until his retort glowed red hot. Suddenly something strange began to happen. Glowing fumes filled the vessel and from the end of the retort dripped a shining liquid that burst into flames. . . .

That Hamburg alchemist was Hennig Brandt, a rather pompous man who insisted on being called *Herr Doktor* Brandt. His first thought was that he had at last found what he had been searching for. Surely this wondrous new material was the philosopher's stone? If so, he had better keep the secret to himself while he made his fortune. And so for six long years he hid his discovery from the world, until his wealth was all but spent. Magical though it was, his phosphorus stubbornly refused to bring the riches of which he dreamed.

Today we know his was a vain hope, but the luminescent material he had made was to create great wealth for a few in the centuries ahead. It was to create untold misery for many more.

products and transported to agricultural areas all over the world. There they are incorporated into the soil, either directly as fertilizers or indirectly as excess phosphorus in manure resulting from the use of high-phosphate animal feeds. Poor land-use practices further increase erosion of this phosphorus-laden soil.

Human actions thus accelerate the natural cycle at two key points: in the entry of phosphorus into the biosphere from rock, and in the movement from soil into aquatic ecosystems. Additionally, by moving phosphorus away from certain spots on the Earth's surface (those where it is mined) and to others (primarily where it is used as fertilizer and animal feed), we radically alter the distribution of this element on the planet's surface. The effect of shifting large quantities to places where it would not naturally be found in high concentration has become a growing concern to aquatic ecologists and others who care about maintaining supplies of clean fresh water.

Most phosphorus moves downhill attached to eroded soil particles—whether over the ground as muddy runoff or in rain-swollen streams or rivers. As people increase the amount of phosphorus in the soil through use of fertilizers, the amount of phosphorus carried downhill per kilogram of soil also increases. The higher the concentration in the soil at the outset, the more is available to release downhill. And although it is likely that less than 5 percent of the phosphorus used as fertilizer in temperate areas makes its way into aquatic ecosystems each year, that is enough to cause major changes in those environments.

Understanding what is happening in the upland soil provides a window on the future of downhill water bodies. If we study soil uphill from lakes and rivers, we know more about the possibilities for the future of those same lakes, rivers, and the estuaries they flow into. Studying changes in the phosphorus cycle and upland soils on a global scale can help us know where to expect problems in the future and may help us reduce excess phosphorus before it is a problem.

MESSING WITH EUTROPHICATION

Under natural conditions, eutrophication can be a centuries-long aging process for some lakes. When human caused (scientists call this "cultural" eutrophication), it can happen in a few years. In many cases, cultural eutrophication can be reversed by greatly reducing the amount of phosphorus entering the water.

Phosphorus is just one of many nutrients that plants need to survive and grow. However, it is particularly important to lake systems because it is the limiting nutrient. In other words, even when plants have all other nutrients in sufficient quantities, they will often lack phosphorus. When phosphorus is added to lake water, plants may suddenly grow very rapidly because all the necessary nutrients are now present. But the excessive growth of one organism may mean the death of another. When thick blooms of algal growth block sunlight from reaching the plants below, the decay of dead algae uses up the available oxygen in the water, suffocating fish and sometimes causing whole populations of species to be lost. Eutrophication not only makes a lake look and smell putrid; it also substantially changes the way the aquatic ecosystem works. It can change the plant community, the food web, and the chemistry of a lake beyond recognition.

Although scientists have been studying eutrophication since the turn of the twentieth century, cultural eutrophication was not recognized as a widespread international problem until the 1950s and '60s, when it became a matter of growing concern in both North America's Great Lakes region and in much of Europe. "Lakes became green and smelly, devoid of fish, unfit as sources of drinking water and unimaginable as places of recreation," writes John Emsley in *The Thirteenth Element: A Sordid Tale of Murder, Fire, and Phosphorus.* Scientists did not always know that phosphorus was the culprit in situations of cultural eutrophication. But in the late 1960s, a research team headed by David Schindler, a professor of Ecology at the University of Alberta in Edmonton, Canada, conducted a revealing experiment on a lake—known as "Lake 227"—in northwestern Ontario. The team divided the lake into two halves. They enriched one half with nitrogen and carbon, and the other half with phosphorus and carbon. The phosphate-enriched basin rapidly became eutrophic, while the other basin remained unchanged.

Phosphorus became widely recognized as the source of fresh water eutrophication problems, and efforts were begun to reduce phosphorus inputs to aquatic ecosystems. In most of the developed world, sewage was diverted around lakes, and most "point" sources (direct outflows of phosphorus-rich effluents from specific farms or processing plants) were cut off.

By the 1990s, phosphates had been removed from most detergents, as well. Yet eutrophication persisted, in part because lakes are efficient recyclers of phosphorus. Under conditions of low oxygen at the lake bottom, phosphorus lying in the sediment reenters the water column and is once again available to be used by algae. The problem also remained because scientists had overlooked another source of phosphorus entering lakes— the non-point source runoff from surrounding lands—which proves much more difficult to control. Currently, non-point runoff from uplands is the main source of phosphorus to most aquatic systems in the developed world.

Worldwide, eutrophication in lakes, rivers, and estuaries appears to be increasing. In the Gulf of Mexico, there is now a large hypoxic zone, or "dead" zone, where the low oxygen content of the water has led to massive die-offs in ocean species. The cause has been traced to nutrient runoff from the grain-growing states of the midwestern United States, carried to the Gulf by the Mississippi River. In New Zealand, an increase in dairy farming and fertilizer use has worsened nutrient pollution in hundreds of shallow lakes and streams. By 1994, the most recent date for which we have worldwide data, significant eutrophication problems were being reported in 54 percent of all lakes and reservoirs in Southeast Asia, 53 percent of those in Europe, 48 percent in North America, 41 percent in South America, and 28 percent in Africa.

The growing recognition that eutrophication is more widespread than we might have initially imagined led us to look for a larger-scale perspective on the issue and its causes. We knew that the immediate cause is human impact on phosphorus cycling: increased phosphorus in soils and increase erosion causes increased erosion to lakes. However, we wanted to understand the eutrophication problem from a global perspective because we believed that this perspective might lead to more effective long-term policies to reduce phosphorus load to aquatic ecosystems. Studying phosphorus increase in soils, as opposed to waiting until fish begin dying off in lakes, can be an effective preventive measure. So far, the systematic assessment of phosphorus in upland soils has not become a standard of the eutrophication management process.

MANAGEMENT: THE GLOBAL PICTURE

Despite widespread controlling of point source pollution, treatment of sewage, and elimination of phosphates from most soaps and detergents, eutrophication continues to worsen as a result of human activity worldwide. What to do?

The phosphorus in agricultural soils could take decades to draw down by reducing use of fertilizers. During that time, changes in farm practices, urban expansion, or climate change could accelerate erosion and—despite the lower input of phosphorus to soil—increase the rate at which phosphorus moves from the soil into aquatic ecosystems. By the time authorities see enough impairment of lakes to want to take action upstream, the fate of the lakes downstream may already have been sealed. By the same logic, however, the long time lags associated with soil phosphorus buildup also mean that action now can prevent expensive and persistent eutrophication problems in the future.

Efforts to control phosphorus runoff have increased dramatically in recent years. Still, most policies and regulations have approached such runoff as a problem of the particular lake, river reach, or estuary, rather than as part of a larger pattern. These efforts have generally involved reducing nearby fertilizer and manure use, limiting erosion, or removing algae from the water directly. In Lake Mendota, the city regularly runs a floating lawn mower–like machine that strips algae and other aquatic plants from the water.

To implement the long-term solution by reducing the phosphorus stored in upland soil is more difficult. Because upland runoff originates from more dispersed sources, it is very difficult to pinpoint the landowners who are responsible. One consequence is that in the United States, programs are usually voluntary rather than mandatory. Effectiveness is limited,

How We Studied the Global Cycle

Although there is not much phosphorus on Earth (only about 0.1 percent of the Earth's crust is phosphorus), it is an essential nutrient for all life forms on our planet. It is found primarily in solid or liquid state and it plays a key role in DNA (the genetic material of most life) and the energy-releasing molecule ATP.

Phosphorus is mined in only a few countries, including the United States, the former Soviet Union, Morocco, and China; however, phosphorus-containing products are consumed in most countries. About 90 percent of the phosphorus mined annually is used to produce fertilizers and animal feeds.

Globally, we knew that by mining phosphorus in order to put it on farms and lawns, we were likely to be creating new areas of high concentration. However, scientists knew little about where accumulation might be most severe. Answering this question would help managers prepare for future eutrophication problems.

We made a budget (analogous to a household's financial budget) to find out how much phosphorus is accumulating in the soil compartment of the global model each year. We wanted to get a largescale picture because we thought we might have more options for management if eutrophication was recognized as a global, rather than just local, problem.

We calculated phosphorus accumulation in soils by determining how much phosphorus enters soils each year and how much is removed. We found values for the amount of mined phosphorus especially if the incentives for participation are not great enough. In some cases, local water authorities attempt to help farmers (and others) to limit phosphorus use or reduce erosion by offering cost sharing from the government. For example, if manure storage pits are needed, the government may offer to pay a share of the cost of creating a manure storage pit.

In other countries such as the Netherlands, restrictions on phosphorus use are more severe and enforced by law rather

that is used for animal feeds, fertilizer, and other easily biodegradable products annually, and added that to the amount of naturally weathered phosphorus. These are the inputs to Earth's soil. We also calculated how much was leaving Earth's soils each year, by estimating how much phosphorus moved into the ocean and other aquatic ecosystems. The difference between the inputs and the outputs is the amount that accumulates in soils on an annual basis, which we estimated to be between 10.5 and 15.5 million metric tons per year.

We completed a similar calculation for soil accumulation of phosphorus before widespread human impact. In the millennia before civilization began, of course, there was no mining of phosphorus—only natural weathering, which released between 1 and 6 million metric tons per year.

By making separate calculations for agricultural areas and nonagricultural, we found that although agricultural areas occupy only a small fraction of the Earth's surface (about 11 percent), most of the phosphorus accumulation occurs there. Year by year, phosphorus continues to build up in the world's farmland.

Separating calculations for poor and rich countries, we found that while poor countries once had a net loss of phosphorus, they now account for more accumulation than rich countries do. Lakes in these countries are probably becoming more susceptible to eutrophication, which is likely to worsen the quality of life of these countries in the future.

The Human Impact on the Phosphorus Cycle

The global phosphorus cycle is made up of four "compartments" of the Earth (crust, soil, freshwater systems, and the ocean), through which—due to the slow process of rock formation in the ocean bottom, subsequent tectonic upheaval, and weathering of rock into soil—it takes millions of years to complete.

Along the way, an atom of phosphorus may go through many more rapid, biological, cycles as it is incorporated into animals or plants in the soil, aquatic, or ocean compartments.

Human activity, however, has radically accelerated some parts of the movement. Mining phosphorus for fertilizer speeds up the release from crust to soil, and clearing land speeds up the movement from soil to aquatic systems.

The result has been a more rapid buildup of phosphorus in soils, and a more widespread eutrophication of lakes and rivers, than occurred in the past.

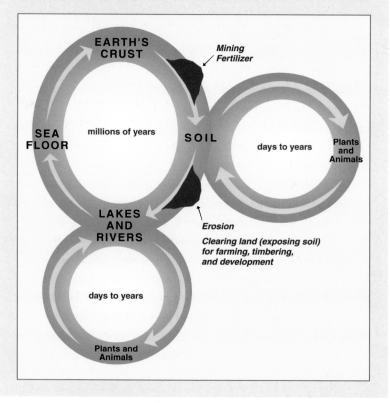

than enacted voluntarily. Eutrophication is a major concern for the Netherlands, due to the country's high population density and intensive agriculture—combined with the fact that the Rhine River water that flows into the Netherlands from other European countries is already high in phosphorus content. Therefore, Dutch farmers are subject to manure quotas: for each acre of land, they may only spread a certain amount of manure. (The amount allowed per acre often depends on the location, size, and type of farm.) For any manure used beyond that amount, the farmer must pay the costs of removal and processing.

As we begin to look at eutrophication as a global phenomenon, we notice that around the world, water subsidizes agricultural systems. Because we do not account for the cost of water pollution when we add up the cost of agricultural production, food production seems cheaper than it actually is. In economic terms, the cost of water pollution caused by agricultural production is an externality to the agricultural production system. If this worked in the past because clean fresh water seemed infinite, it no longer makes sense to take fresh water supply for granted. Since both food and water are necessary for human survival, it might appear that we are facing a difficult choice between producing food and protecting aquatic resources. However, just as we cannot afford to ignore water pollution, we cannot afford to solve the problem simply by reducing agricultural production.

The growing human population and increasing demand for food, along with the lack of vacant arable land for use in agricultural expansion, mean that agricultural production will probably have to become more intense and efficient on land already in production.

Win-win solutions would change regulations and incentives to bring the costs and benefits of agricultural production and water quality protection into better balance. For example, governments could create markets for nitrogen and phosphorus runoff.

Each farm would be allotted a certain amount of permissible pollution or runoff. If the farmer could find a way to emit less pollution, he could then sell some of his pollution rights to another farm. If set up correctly, the market would quickly adjust to the value of water—thereby internalizing the cost of water pollution.

Alternatively, governments could tax fertilizer use to create an incentive for conservation, using the tax revenues to improve water quality.

Such methods can prove economically as well as environmentally beneficial. A few years ago, New York City, which gets its water supply from the Catskill mountain area, found that its water was no longer meeting EPA [Environmental Protection Agency] standards due to fertilizers and pesticides in the soil. City officials realized they would have to spend between $6 billion and $8 billion to build a water filtration plant to clean the water. Instead, they were able to purchase watershed land in the Catskills sufficient to naturally filter the water for only $1.5 billion. Quality of the water supply was increased and money was saved.

Comparable solutions may be feasible at national and global scales, making it possible to conserve fertilizer, stabilize soils, and improve water quality while reducing costs. Effective policies might include the establishment of national or international phosphorus markets and better tracking of phosphorus around the world, in order to ensure that some areas are not being overwhelmed by excess nutrients. Phosphorus markets could ensure that products such as manure are moved to where they can be used.

What Happens When People Take Water From Our National Wildlife Refuges?

Without water, there is no life. And there is a fixed amount of water on the planet. So it is not surprising that with more than 6 billion people living on the Earth, competition over who gets to use the available water can become intense. Some countries fight wars over water. In the United States, we fight legal battles.

Consider, for example, the case of Tule Lake and Lower Klamath National Wildlife Refuges, located on the California-Oregon border. The seeds of the problem were sown in the early 1900s, when the government invited farmers to move to the area to begin "agricultural reclamation." Just as in the Everglades, that land of great biological diversity, the government encouraged people to drain this western land for farming. Years of diverting water from rivers and wetlands has taken its toll. The area is a flyway for 80% of the western migrating birds. Because of changes to the water system, the birds' numbers have dropped from 6 million to 1.5 million. Fish species have also become endangered.

In 1999, the U.S. Fish and Wildlife Service announced that agriculture, not the marshes, would receive the water of the Tule Lake and Lower Klamath area. But in 2001, the governors of both California and Oregon declared that because of an ongoing drought, no water from the Upper Klamath Lake would be available for irrigation. In 2002, 10 conservation groups joined with tribal groups and commercial fisherman to challenge the federal government's policy of giving the water to farms.

The following article by journalist Michael Statchell, "Troubled Waters," introduces some of the problems of water use and misuse in our many national wildlife refuges. In 1903, President Theodore Roosevelt had the foresight to recognize that special lands needed protection and established the first National Wildlife Refuge. Unfortunately, land without water quickly loses its environmental, social, and economic value. And water that feeds protected areas such as wildlife refuges can be diverted or become so polluted that the health of the refuge is endangered.

Today, there are refuges where people are solving the problem through river reclamation projects and working with neighboring landowners. About the progress in the Big Muddy National Fish and Wildlife Refuge in Missouri, Jim Kurth, deputy chief of the National Wildlife Refuge System, says, "With flooding and channel restoration, we're letting the river behave like a river." The challenge facing U.S. policymakers is to allow this to happen in all National Wildlife Refuges.

—The Editor

Troubled Waters
by Michael Statchell

The view from the levee encompasses a scene of beauty both bucolic and bizarre. Off to the right, and well below the dirt road that splits the viewscape almost to the horizon, lie fields of emerald green alfalfa undulating in a gentle breeze. To the left and higher than the road stretches a vast flat plain of brilliant yellow that, from a distance, could be a field of ripening flax. But closer examination reveals that this "crop" is actually a giant pool of standing water covered by a thick mat of oxygen-gobbling algae.

To Phil Norton, recently retired manager of the six Klamath Basin national wildlife refuges on the Oregon-California border, this vista at Tule Lake refuge is an all-too-familiar sign of an ecosystem being slowly destroyed by lack of fresh water. More than 60 farmers who lease 20,500 acres [8,296 hectares] on the refuge (compared with 14,000 acres [5,666 hectares] devoted to wetlands habitat) siphon off much of this precious resource, thanks to a 1964 law that they say gives them rights to the cropland.

To replenish refuge ponds and vary their levels at different times of year—necessary to maintain appropriate wildlife habitat—Tule Lake's staff needs large, regular inflows of fresh water. Yet the refuges get only what little water is left after the federal Bureau of Reclamation supplies farmers and other competing users. Even in wet years, the Klamath refuges do not get as much water as they need. And during times of drought, battles over this critical resource turn ugly, with the protected areas generally coming out the losers.

Water is the lifeblood of the National Wildlife Refuge System, and inadequate water quantity and quality are among the biggest threats to refuges nationwide. "On the vast majority of our national wildlife refuges, water is the critical issue," says the system's deputy chief, Jim Kurth. "It is the resource that limits our ability to provide first-class fish and wildlife habitat." Short water supply is the most pervasive problem, with managers from at least 150 refuges reporting conflicts with competing users. Other challenges include flows tainted by pesticides, fertilizers, animal waste and other pollutants; huge navigation and flood control projects that alter refuge water sources; inundation of saltwater marshes by rising seas; and wetlands draining by private land owners. Such problems are only expected to worsen in the future, adds Kurth, "as population grows, and there is increasing demand for water from urban areas, agriculture and industry."

At Tule Lake, the effects of water shortages are obvious. Instead of seasonal wetlands, deep pools, wading shallows and vegetation-rich islands, there are muck-bottomed ponds containing fertilizer and other agricultural chemicals where a few endangered sucker fish struggle to survive. The water's surface is covered by a carpet of algae thick enough in places for waterfowl to walk on.

The impact on wildlife has been predictably dramatic. Strategically located on the Pacific Flyway, Klamath's wetlands historically served as a critical stopover for 80 percent of North American migratory waterfowl that travel this western route. A half century ago, some 6 million ducks, geese and swans would shade out the sun twice a year as they migrated between their northern breeding grounds and southern wintering habitat. But today those epochal wildlife spectacles are a distant memory, and the migration is down to about 1.5 million birds.

"The importance of the Klamath Basin to the Pacific Flyway cannot be overstated," says NWF [National Wildlife Federation] attorney Jan Hasselman, who before joining the federation in

2001 represented both conservation groups and commercial fishermen in court battles over the region's water. "Until we have a fair plan to allocate the Klamath Basin's scarce water over the long term, these crown jewels of our wildlife refuge system will continue to deteriorate."

Today's troubles on the refuges have been more than 100 years in the making. At the turn of the century, the Klamath Basin was covered by some 190,000 acres [76,890 hectares] of shallow wetlands, an area larger than Lake Tahoe. The Klamath River and its tributaries were among the richest salmon and steelhead trout habitats in the West, and lakes teemed with freshwater mullet (also called sucker fish). But in 1902, the newly formed Bureau of Reclamation began to build dams, reservoirs and pipelines, diverting water from the river system and draining 80 percent of its wetlands to uncover rich peat soils for farming. Both Tule Lake and Lower Klamath refuges also were created during this period, but because of the federal government's interpretation of refuge water rights—an interpretation conservationists reject—the sanctuaries have been on virtual life support ever since.

In 2001, decades of competition for the basin's scarce water reached an acrimonious climax when a drought left reservoirs so low that the Bureau of Reclamation, acting under court order, cut off two-thirds of the water routinely delivered to farmers and sent it instead into lakes and rivers harboring endangered salmon and mullet. Waterfowl habitat on the refuges, meanwhile, did not receive much more than a tea cup. Two years later, water users remain locked in a two-sided battle—with conservationists, Indian tribes and commercial fishermen advocating that more water be left in the river system and its wetlands, and farmers trying to take more out. Last fall, NWF entered the fray when it joined a coalition of ten conservation groups that have filed a lawsuit challenging the government's policy of favoring farmers over wildlife on the Klamath refuges.

Elsewhere, water woes vary from refuge to refuge, yet they are common in nearly every region of the country. Some of the biggest threats come from mega-projects sponsored by the U.S. Army Corps of Engineers. Stretching for 261 riverine miles [420 km] through Illinois, Iowa, Wisconsin and Minnesota, the Upper Mississippi River Wildlife and Fish Refuge provides a typical example. A Corps proposal to expand six locks on the Mississippi downstream from the refuge threatens the fecund ecosystems of its mainstream river, floodplain and wetlands—critical migratory bird stopovers on the Mississippi Flyway. The refuge harbors some 300 avian species, including 40 percent of the continent's waterfowl population, as well as 100 species of fish and 40 species of freshwater mussels.

According to Upper Mississippi refuge biologist Eric Nelson, bigger locks would encourage more barge traffic and require increased channel maintenance, exacerbating adverse effects of navigation already being felt on the refuge. "We fear more erosion and siltation, which will mean the loss of island habitat crucial to protecting backwaters from wind and wave action," says Nelson. "This will affect everything from migrating waterfowl to fisheries."

Another Corps project to facilitate barge traffic—this one to help farmers get grain to market faster in Arkansas—portends major problems for the state's White River and Cache River refuges. Together, these refuges protect more than 200,000 acres [80,937 hectares] of bottomland hardwood forest and support North America's largest population of wintering mallards, along with bald eagles, least terns and Arkansas' most important black bear population. The Corps plans to dredge or dike nearly 100 miles [161 km] running through the White River preserve, which would increase siltation and erosion as well as damage rare and sensitive forested wetlands on both refuges. "It's like building a four-lane highway just to deliver the mail," complains White River refuge manager Larry Mallard. Both refuges also are threatened by a number of irrigation projects, most notably the Grand Prairie Demonstration Project, which would siphon off more than a billion gallons of water a day.

On scores of other refuges, water pollution is the biggest problem, sometimes killing vast numbers of fish and birds.

Located on the state's largest inland body of water, California's Sonny Bono Salton Sea National Wildlife Refuge is perhaps the most dramatic example. Each day, the sea absorbs roughly the equivalent of a one-mile-long trainload of contaminants, including pesticides, mineral salts and fertilizers such as nitrogen and phosphorus. From the north, the effluent seeps down as agricultural runoff from farming operations in the Imperial Valley. From the south, the New River, reputedly the filthiest waterway in the United States, flows out of Mexico and dumps industrial effluent, raw sewage, slaughterhouse wastes and other toxins into this contaminated crucible. And because the sea is a sump with no outlets, expanding and contracting with cycles of inflowing water and evaporation, concentrations of pollutants only increase over time.

The effects on refuge wildlife have been disastrous. So many birds die after eating fish poisoned in the sea's chemical soup that the refuge operates two incinerators to dispose of their carcasses, as well as the thousands of fish that routinely float bellyup in the stagnant water. In the summer of 1996, one outbreak of avian botulism alone killed over 8,500 white pelicans and more than 1,125 endangered brown pelicans.

At South Carolina's Cape Romaine National Wildlife Refuge, the primary water contaminant is shrimp bait. Between 1988 and 1997, the number of shrimpers using the refuge rose from about 100 to a staggering 20,000, all of them baiting with unsterilized fish meal mixed with clay or cement, a combination that causes high nutrient levels lethal to aquatic life. A massive die-off of oysters in the refuge, for example, has triggered a sharp decline in its American oystercatcher population.

But the water news on refuges is not all bad. Deputy Chief Kurth points to an encouraging trend toward buying private land in areas prone to flooding and creating new wildlife sanctuaries there. As an example, he points to the Big Muddy National Fish and Wildlife Refuge in Missouri, which was established in 1994 along Missouri River bottomland that shelters a variety of resident and migratory birds, as well as endangered species such as piping plover, pallid sturgeon and gray bat. "With flooding and channel restoration, we're letting the river behave like a river," says Kurth.

On some refuges, even the Corps of Engineers is playing the hero. In Arizona, for example, the Bill Williams River National Wildlife Refuge boasts one of the biggest and most diverse stretches of riparian habitat on the Lower Colorado River system. A haven for desert bighorn sheep, mule deer, javelina and mountain lions, along with more than 335 species of birds, 34 butterflies and 15 bats, this critical wildlife habitat is at the mercy of Corps officials who operate a dam 40 miles [64 km] upstream. Yet so far, the Corps has provided refuge manager Dick Gilbert enough water to maintain his sanctuary's diversity. "They're excellent neighbors," Gilbert says.

At the threatened Upper Mississippi refuge, the Corps is participating in a major habitat restoration project—even as it proceeds with its controversial lock-and-dam expansion proposal downstream. Created in 1924, the refuge's riverine habitats were drastically altered a decade later when the Corps constructed its initial series of dams and locks, converting a free-flowing river into a series of giant pools between the dams. The changes led to loss of emergent vegetation, which in turn reduced numbers of puddle ducks such as mallards.

Two years ago, refuge managers—working both with the Corps and other federal and state agencies—experimented at one of these artificial lakes in Wisconsin, using the lock-and-dam system to lower water levels by 18 inches [46 cm] during the summer. The benefits to wildlife were swift and dramatic. New areas of dry land quickly sprouted vegetation, and more light penetrating shallow waters increased aquatic plants, improving habitat for both waterfowl and fish. The project turned out to be "a wonderful, innovative way to get enhanced wildlife habitat on the river," says refuge biologist Eric Nelson. "We're going to have a diversity of species we have not seen here in decades."

NO
ADMITTANCE
DO NOT ENTER THE WATER.
NO SWIMMING,
FISHING, WADING.

THE WATER & LAKE
SEDIMENT CONTAIN
HAZARDOUS SUBSTANCES.

What Are the Threats to Groundwater?

Groundwater makes up about 97% of the available (not frozen) freshwater on the planet. But you can't fish in it, you can't swim in it, and you can't wade in it to cool off. For this reason, groundwater is often forgotten. Groundwater is the water that lies between the soil or rock particles under the Earth's surface. It doesn't flow in underground rivers, but there can be large deposits of it, called aquifers. But before you assume that there are endless supplies of groundwater on the planet, consider that the available freshwater makes up only about 1% of all the Earth's water. The rest is in the oceans (97%) or is frozen (the 2% in glaciers).

Besides its location, there are several things that make groundwater different from surface water. Obviously, we cannot see it. So, unlike lakes that get scummy or rivers that turn white with pollution, we may not know for many years when groundwater becomes polluted. Secondly, it moves much differently from surface water. Groundwater movement depends on the lay of the land and the type of soil or rock it is in. Thus, pollutants can move slowly, quickly, or quite unpredictably. Thirdly, groundwater lacks the flushing, or cleaning, action of surface water. Therefore, contaminants can remain underground and undetected for many years. Lastly, groundwater recharges (refills) much more slowly than surface water does. This means that once the supply is depleted, it could take hundreds or even thousands of years for water to refill the spaces in the soil. Groundwater may be out of sight, but it cannot be kept out of mind if we want to have a continued supply of clean drinking and irrigation water.

The following World Watch Institute Paper, "Groundwater Shock," examines groundwater issues from all over the world. World Watch Institute is an independent, nonprofit environmental research organization based in Washington, D.C. World Watch papers give in-depth analysis of issues and are reviewed by experts in the field. Their reports are published in five languages. In this report, author Payla Sampat easily builds the case that threats to groundwater are numerous and serious. Pollution is a huge threat: pollution from landfills, from dumpsites, from agriculture, and from industries. Too much water being pumped from groundwater is another threat. In some places—California and Florida, for example—the land literally sinks, or subsides, when too much groundwater is pumped out. The challenge is for governments and citizens to protect the valuable groundwater resources before they are irretrievably degraded. Sampat aptly ends his article with a quote from Henry David Thoreau: "Heaven is under our feet, as well as over our heads."

—The Editor

Groundwater Shock
by Payla Sampat

The Mississippi River occupies a mythic place in the American imagination, in part because it is so huge. At any given moment, on average, about 2,100 billion liters of water are flowing across the Big Muddy's broad bottom. If you were to dive about 35 feet down and lie on that bottom, you might feel a sense of awe that the whole river was on top of you. But in one very important sense, you'd be completely wrong. At any point in time, only 1 percent of the water in the Mississippi River system is in the part of the river that flows downstream to the Gulf of Mexico. The other 99 percent lies beneath the bottom, locked in massive strata of rock and sand.

This is a distinction of enormous consequence. The availability of clean water has come to be recognized as perhaps the most critical of all human security issues facing the world in the next quarter-century—and what is happening to water buried under the bottoms of rivers, or under our feet, is vastly different from what happens to the "surface" water of rivers, lakes, and streams. New research finds that contrary to popular belief, it is groundwater that is most dangerously threatened. Moreover, the Mississippi is not unique in its ratio of surface to underground water; worldwide, 97 percent of the planet's liquid freshwater is stored in aquifers.

In the early centuries of civilization, surface water was the only source we needed to know about. Human population was less than a tenth of one percent the size it is now; settlements were on river banks; and the water was relatively clean. We still think of surface water as being the main resource. So it's easy to think that the problem of contamination is mainly one of surface water: it is polluted rivers and streams that threaten health in times of flood, and that have made waterborne diseases a major killer of humankind. But in the past century, as population has almost quadrupled and rivers have become more depleted and polluted, our dependence on pumping groundwater has soared—and as it has, we've made a terrible discovery. Contrary to the popular impression that at least the waters from our springs and wells are pure, we're uncovering a pattern of pervasive pollution there too. And in these sources, unlike rivers, the pollution is generally irreversible.

This is largely the work of another hidden factor: the rate of groundwater renewal is very slow in comparison with that of surface water. It's true that some aquifers recharge fairly quickly, but the average recycling time for groundwater is 1,400 years, as opposed to only 20 days for river water. So when we pump out groundwater, we're effectively removing it from aquifers for generations to come. It may evaporate and return to the atmosphere quickly enough, but the resulting rainfall (most of which falls back into the oceans) may take centuries to recharge the aquifers once they've been depleted. And because water in aquifers moves through the Earth with glacial slowness, its pollutants continue to accumulate.

Unlike rivers, which flush themselves into the oceans, aquifers become sinks for pollutants, decade after decade—thus further diminishing the amount of clean water they can yield for human use.

Perhaps the largest misconception being exploded by the spreading water crisis is the assumption that the ground we stand on—and what lies beneath it—is solid, unchanging, and inert. Just as the advent of climate change has awakened us to the fact that the air over our heads is an arena of enormous forces in the midst of titanic shifts, the water crisis has revealed that, slow-moving though it may be, groundwater is part of a system of powerful hydrological interactions—between earth, surface water, sky, and sea—that we ignore at our peril. A few years ago, reflecting on how human activity is beginning to affect climate, Columbia University scientist Wallace Broecker warned, "The climate system is an angry beast and we are poking it with sticks." A similar statement might now be made about the system under our feet. If we continue to drill holes into it—expecting it to swallow our waste and yield freshwater in return—we may be toying with an outcome no one could wish.

VALUING GROUNDWATER

For most of human history, groundwater was tapped mainly in arid regions where surface water was in short supply. From Egypt to Iran, ancient Middle Eastern civilizations used periscope-like conduits to funnel spring water from mountain slopes to nearby towns—a technology that allowed settlement to spread out from the major rivers. Over the centuries, as populations and cropland expanded, innovative well-digging techniques evolved in China, India, and Europe. Water became such a valuable resource that some cultures developed elaborate mythologies imbuing underground water and its seekers with special powers. In medieval Europe, people called water witches or dowsers were believed to be able to detect groundwater using a forked stick and mystical insight.

In the second half of the 20th century, the soaring demand for water turned the dowsers' modernday counterparts into a major industry. Today, major aquifers are tapped on every continent, and groundwater is the primary source of drinking water for more than 1.5 billion people worldwide. The aquifer that lies beneath the Huang-Huai-Hai plain in eastern China alone supplies drinking water to nearly 160 million people. Asia as a whole relies on its groundwater for nearly one-third of its drinking water supply. Some of the largest cities in the developing world—Jakarta, Dhaka, Lima, and Mexico

City, among them—depend on aquifers for almost all their water. And in rural areas, where centralized water supply systems are undeveloped, groundwater is typically the sole source of water.

More than 95 percent of the rural U.S. population depends on groundwater for drinking. A principal reason for the explosive rise in groundwater use since 1950 has been a dramatic expansion in irrigated agriculture. In India, the leading country in total irrigated area and the world's third largest grain producer, the number of shallow tubewells used to draw groundwater surged from 3,000 in 1960 to 6 million in 1990. While India doubled the amount of its land irrigated by surface water between 1950 and 1985, it increased the area watered by aquifers 113-fold. Today, aquifers supply water to more than half of India's irrigated land. The United States, with the third highest irrigated area in the world, uses groundwater for 43 percent of its irrigated farmland.

Worldwide, irrigation is by far the biggest drain on freshwater: it accounts for about 70 percent of the water we draw from rivers and wells each year.

Other industries have been expanding their water use even faster than agriculture—and generating much higher profits in the process. On average, a ton of water used in industry generates roughly $14,000 worth of output—about 70 times as much profit as the same amount of water used to grow grain. Thus, as the world has industrialized, substantial amounts of water have been shifted from farms to more lucrative factories. Industry's share of total consumption has reached 19 percent and is likely to continue rising rapidly. The amount of water available for drinking is thus constrained not only by a limited resource base, but by competition with other, more powerful users.

And as rivers and lakes are stretched to their limits—many of them dammed, dried up, or polluted—we're growing more and more dependent on groundwater for all these uses. In Taiwan, for example, the share of water supplied by groundwater almost doubled from 21 percent in 1983 to over

40 percent in 1991. And Bangladesh, which was once almost entirely river- and stream-dependent, dug over a million wells in the 1970s to substitute for its badly polluted surface-water supply. Today, almost 90 percent of its people use only groundwater for drinking.

Even as our dependence on groundwater increases, the availability of the resource is becoming more limited. On almost every continent, many major aquifers are being drained faster than their natural rate of recharge. Groundwater depletion is most severe in parts of India, China, the United States, North Africa, and the Middle East. Under certain geological conditions, groundwater overdraft can cause aquifer sediments to compact, permanently shrinking the aquifer's storage capacity. This loss can be quite considerable, and irreversible. The amount of water storage capacity lost because of aquifer compaction in California's Central Valley, for example, is equal to more than 40 percent of the combined storage capacity of all human-made reservoirs across the state.

As the competition among factories, farms, and households intensifies, it's easy to overlook the extent to which freshwater is also required for essential ecological services. It is not just rainfall, but groundwater welling up from beneath, that replenishes rivers, lakes, and streams. In a study of 54 streams in different parts of the country, the U.S. Geological Survey (USGS) found that groundwater is the source for more than half the flow, on average. The 492 billion gallons (1.86 cubic kilometers) of water aquifers add to U.S. surface water bodies each day is nearly equal to the daily flow of the Mississippi. Groundwater provides the base contribution for the Mississippi, the Niger, the Yangtze, and many more of the world's great rivers—some of which would otherwise not be flowing year-round. Wetlands, important habitat for birds, fish, and other wildlife, are often largely groundwaterfed, created in places where the water table overflows to the surface on a constant basis. And while providing surface bodies with enough water to keep them stable, aquifers also help prevent them from flooding: when it rains heavily, aquifers beneath rivers soak up the excess water, preventing the surface flow from rising too rapidly and overflowing onto neighboring fields and towns. In tropical Asia, where the hot season can last as long as 9 months, and where monsoon rains can be very intense, this dual hydrological service is of critical value.

Numerous studies have tracked the extent to which our increasing demand on water has made it a resource critical to a degree that even gold and oil have never been. It's the most valuable thing on Earth. Yet, ironically, it's the thing most consistently overlooked, and most widely used as a final resting place for our waste. And, of course, as contamination spreads, the supplies of usable water get tighter still.

TRACKING THE HIDDEN CRISIS
In 1940, during the Second World War, the U.S. Department of the Army acquired 70 square kilometers [27 square miles] of land around Weldon Spring and its neighboring towns near St. Louis, Missouri. Where farmhouses and barns had been, the Army established the world's largest TNT-producing facility. In this sprawling warren of plants, toluene (a component of gasoline) was treated with nitric acid to produce more than a million tons of the explosive compound each day when production was at its peak.

Part of the manufacturing process involved purifying the TNT—washing off unwanted "nitroaromatic" compounds left behind by the chemical reaction between the toluene and nitric acid. Over the years, millions of gallons of this red-colored muck were generated. Some of it was treated at wastewater plants, but much of it ran off from the leaky treatment facilities into ditches and ravines, and soaked into the ground. In 1945, when the Army left the site, soldiers burned down the contaminated buildings but left the red-tinged soil and the rest of the site as they were. For decades, the site remained abandoned and unused.

Then, in 1980, the U.S. Environmental Protection Agency (EPA) launched its "Superfund" program, which required the cleaning up of several sites in the country that were contaminated with hazardous waste. Weldon Spring made it to the list of sites that were the highest priority for cleanup. The Army Corps of Engineers was assigned the task, but what the Corps workers found baffled them. They expected the soil and vegetation around the site to be contaminated with the nitroaromatic wastes that had been discarded there. When they tested the groundwater, however, they found that the chemicals were showing up in people's wells, in towns several miles from the site—a possibility that no one had anticipated, because the original pollution had been completely localized. Geologists determined that there was an enormous plume of contamination in the water below the TNT factory—a plume that over the previous 35 years had flowed through fissures in the limestone rock to other parts of the aquifer.

The Weldon Spring story may sound like an exceptional case of clumsy planning combined with a particularly vulnerable geological structure. But in fact there is nothing exceptional about it all. Across the United States, as well as in parts of Europe, Asia, and Latin America, human activities are sending massive quantities of chemicals and pollutants into groundwater. This isn't entirely new, of course; the subterranean world has always been a receptacle for whatever we need to dispose of—whether our sewage, our garbage, or our dead. But the enormous volumes of waste we now send underground, and the deadly mixes of chemicals involved, have created problems never before imagined.

What Weldon Spring shows is that we can't always anticipate where the pollution is going to turn up in our water, or how long it will be from the time it was deposited until it reappears. Because groundwater typically moves very slowly—at a speed of less than a foot a day, in some cases—damage done to aquifers may not show up for decades. In many parts of the world, we are only just beginning to discover contamination caused by practices of 30 or 40 years ago.

Some of the most egregious cases of aquifer contamination now being unearthed date back to Cold War era nuclear testing and weapons-making, for example.

And once it gets into groundwater, the pollution usually persists: the enormous volume, inaccessibility, and slow rate at which groundwater moves make aquifers virtually impossible to purify.

As this covert crisis unfolds, we are barely beginning to understand its dimensions. Few countries track the health of their aquifers—their enormous size and remoteness make them extremely expensive to monitor. As the new century begins, even hydrogeologists and health officials have only a hazy impression of the likely extent of groundwater damage in different parts of the world. . . .

THE FILTER THAT FAILED: PESTICIDES IN YOUR WATER

Pesticides are designed to kill. The first synthetic pesticides were introduced in the 1940s, but it took several decades of increasingly heavy use before it became apparent that these chemicals were injuring non-target organisms—including humans. One reason for the delay was that some groups of pesticides, such as organochlorines, usually have little effect until they bioaccumulate. Their concentration in living tissue increases as they move up the food chain. So eventually, the top predators—birds of prey, for example—may end up carrying a disproportionately high burden of the toxin. But bioaccumulation takes time, and it may take still more time before the effects are discovered. In cases where reproductive systems are affected, the aftermath of this chemical accumulation may not show up for a generation.

Even when the health concerns of some pesticides were recognized in the 1960s, it was easily assumed that the real dangers lay in the dispersal of these chemicals among animals and plants—not deep underground.

It was assumed that very little pesticide would leach below the upper layers of soil, and that if it did, it would be degraded before it could get any deeper. Soil, after all, is known to be a natural filter, which purifies water as it trickles through. It was thought that industrial or agricultural chemicals, like such natural contaminants as rock dust, or leaf mold, would be filtered out as the water percolated through the soil.

But over the past 35 years, this seemingly safe assumption has proved mistaken. Cases of extensive pesticide contamination of groundwater have come to light in farming regions of the United States, Western Europe, Latin America, and South Asia.

What we now know is that pesticides not only leach into aquifers, but sometimes remain there long after the chemical is no longer used. DDT, for instance, is still found in U.S. waters even though its use was banned 30 years ago. In the San Joaquin Valley of California, the soil fumigant DBCP (dibromochloropropane), which was used intensively in fruit orchards before it was banned in 1977, still lurks in the region's water supplies. Of 4,507 wells sampled by the USGS between 1971 and 1988, nearly a third had DBCP levels that were at least 10 times higher than allowed by the current drinking water standard.

In places where organochlorines are still widely used, the risks continue to mount. After half a century of spraying in the eastern Indian states of West Bengal and Bihar, for example, the Central Pollution Control Board found DDT in groundwater at levels as high as 4,500 micrograms per liter—several thousand times higher than what is considered a safe dose.

The amount of chemical that reaches groundwater depends on the amount used above ground, the geology of the region, and the characteristics of the pesticide itself. In some parts of the midwestern United States, for example, although pesticides are used intensively, the impermeable soils of the region make it difficult for the chemicals to percolate underground. The fissured aquifers of southern Arizona, Florida, Maine, and southern California, on the other hand, are very vulnerable to pollution—and these too are places where pesticides are applied in large quantities.

Pesticides are often found in combination, because most farms use a range of toxins to destroy different kinds of insects, fungi, and plant diseases. The USGS detected two or more pesticides in groundwater at nearly a quarter of the sites sampled in its National Water Quality Assessment between 1993 and 1995. In the Central Columbia Plateau aquifer, which extends over the states of Washington and Idaho, more than two-thirds of water samples contained multiple pesticides.

Scientists aren't entirely sure what happens when these chemicals and their various metabolites come together. We don't even have standards for the many hundred *individual* pesticides in use—the EPA has drinking water standards for just 33 of these compounds—to say nothing of the infinite variety of toxic blends now trickling into the groundwater.

While the most direct impacts may be on the water we drink, there is also concern about what occurs when the pesticide-laden water below farmland is pumped back up for irrigation. One apparent consequence is a reduction in crop yields. In 1990, the now-defunct U.S. Office of Technology Assessment reported that herbicides in shallow groundwater had the effect of "pruning" crop roots, thereby retarding plant growth.

From Green Revolution to Blue Baby: The Slow Creep of Nitrogen

Since the early 1950s, farmers all over the world have stepped up their use of nitrogen fertilizers. Global fertilizer use has grown ninefold in that time. But the larger doses of nutrients often can't be fully utilized by plants. A study conducted over a 140,000 square kilometer [54,054-square-mile] region of Northern China, for example, found that crops used on average only 40 percent of the nitrogen that was applied. An almost identical degree of waste was found in Sri Lanka. Much of the excess fertilizer dissolves in irrigation water, eventually trickling through the soil into underlying aquifers.

Joining the excess chemical fertilizer from farm crops is the organic waste generated by farm animals, and the sewage produced by cities. Livestock waste forms a particularly potent tributary to the stream of excess nutrients flowing into the environment, because of its enormous volume. In the United States, farm animals produce 130 times as much waste as the country's people do—with the result that millions of tons of cow and pig feces are washed into streams and rivers, and some of the nitrogen they carry ends up in groundwater. To this Augean [extremely difficult and distasteful] burden can be added the innumerable leaks and overflows from urban sewage systems, the fertilizer runoff from suburban lawns, golf courses, and landscaping, and the nitrates leaking (along with other pollutants) from landfills.

There is very little historical information available about trends in the pollution of aquifers. But several studies show that nitrate concentrations have increased as fertilizer applications and population size have grown. In California's San Joaquin–Tulare Valley, for instance, nitrate levels in groundwater increased 2.5 times between the 1950s and 1980s—a period in which fertilizer inputs grew six-fold.

Levels in Danish groundwater have nearly tripled since the 1940s. As with pesticides, the aftermath of this multi-sided assault of excess nutrients has only recently begun to become visible, in part because of the slow speed at which nitrate moves underground.

What happens when nitrates get into drinking water? Consumed in high concentrations—at levels above 10 milligrams (mg) per liter, but usually on the order of 100 mg/liter—they can cause infant methemoglobinemia, or so-called blue-baby syndrome. Because of their low gastric acidity, infant digestive systems convert nitrate to nitrite, which blocks the oxygen-carrying capacity of a baby's blood, causing suffocation and death.

Since 1945, about 3,000 cases have been reported worldwide—nearly half of them in Hungary, where private wells have particularly high concentrations of nitrates. Ruminant livestock such as goats, sheep, and cows, are vulnerable to methemoglobinemia in much the same way infants are, because their digestive systems also quickly convert nitrate to nitrite. Nitrates are also implicated in digestive tract cancers, although the epidemiological link is still uncertain.

In cropland, nitrate pollution of groundwater can have a paradoxical effect. Too much nitrate can weaken plants' immune systems, making them more vulnerable to pests and disease. So when nitrate-laden groundwater is used to irrigate crops that are also being fertilized, the net effect may be to reduce, rather than to increase production. This kind of over-fertilizing makes wheat more susceptible to wheat rust, for example, and it makes pear trees more vulnerable to fire blight.

In assembling studies of groundwater from around the world, we have found that nitrate pollution is pervasive—but has become particularly severe in the places where human population—and the demand for high food productivity—is most concentrated.

In the northern Chinese counties of Beijing, Tianjin, Hebei, and Shandong, nitrate concentrations in groundwater exceeded 50 mg/liter in more than half of the locations studied. (The World Health Organization [WHO] drinking water guideline is 10 mg/liter.) In some places, the concentration had risen as high as 300 mg/liter. Since then, these levels may have increased, as fertilizer applications have escalated since the tests were carried out in 1995 and will likely increase even more as China's population (and demand for food) swells, and as more farmland is lost to urbanization, industrial development, nutrient depletion, and erosion.

Reports from other regions show similar results. The USGS found that about 15 percent of shallow groundwater sampled below agricultural and urban areas in the United States had nitrate concentrations higher than the 10 mg/liter guideline. In Sri Lanka, 79 percent of wells sampled by the British Geological Survey had nitrate levels that exceeded this guideline.

Some 56 percent of wells tested in the Yucatán peninsula in Mexico had levels above 45 mg/liter. And the European Topic Centre on Inland Waters found that in Romania and Moldova, more than 35 percent of the sites sampled had nitrate concentrations higher than 50 mg/liter.

From Tank of Gas to Drinking Glass: The Pervasiveness of Petrochemicals

Drive through any part of the United States, and you'll probably pass more gas stations than schools or churches. As you pull into a station to fill up, it may not occur to you that you're parked over one of the most pervasive threats to groundwater: an underground storage tank (UST) for petroleum.

Many of these tanks were installed two or three decades ago and, having been left in place long past their expected lifetimes, have rusted through in places—allowing a steady leakage of gasoline into the ground. Because they're underground, they're expensive to dig up and repair, so the leakage in some cases continues for years.

Petroleum and its associated chemicals—benzene, toluene, and gasoline additives such as MTBE [methyl tertiary-butyl ether]—constitute the most common category of groundwater contaminant found in aquifers in the United States.

Many of these chemicals are also known or suspected to be cancer-causing. In 1998, the EPA found that over 100,000 commercially owned petroleum USTs were leaking, of which close to 18,000 are known to have contaminated groundwater. In Texas, 223 of 254 counties report leaky USTs, resulting in a silent disaster that, according to the EPA, "has affected, or has the potential to affect, virtually every major and minor aquifer in the state." Household tanks, which store home heating oil, are a problem as well. Although the household tanks aren't subject to the same regulations and inspections as commercial ones, the EPA says they are "undoubtedly leaking."

Outside the United States, the world's ubiquitous petroleum storage tanks are even less monitored, but spot tests suggest that the threat of leakage is omnipresent in the industrialized world. In 1993, petroleum giant Shell reported that a third of its 1,100 gas stations in the United Kingdom were known to have contaminated soil and groundwater.

Another example comes from the eastern Kazakh town of Semipalatinsk, where 6,460 tons of kerosene have collected in an aquifer under a military airport, seriously threatening the region's water supplies.

The widespread presence of petrochemicals in groundwater constitutes a kind of global malignancy, the danger of which has grown unobtrusively because there is such a great distance between cause and effect. An underground tank, for example, may take years to rust; it probably won't begin leaking until long after the people who bought it and installed it have left their jobs. Even after it begins to leak, it may take several more years before appreciable concentrations of chemicals appear in the aquifer—and it will likely be years beyond that before any health effects show up in the local population. By then, the trail may be decades old. So it's quite possible that any cancers occurring today as a result of leaking USTs might originate from tanks that were installed half a century ago. At that time, there were gas tanks sufficient to fuel 53 million cars in the world; today there are enough to fuel almost 10 times that number.

FROM SEDIMENT TO SOLUTE:
THE EMERGING THREAT OF NATURAL CONTAMINANTS

In the early 1990s, several villagers living near India's West Bengal border with Bangladesh began to complain of skin sores that wouldn't go away. A researcher at Calcutta's Jadavpur University, Dipankar Chakraborti, recognized the lesions immediately as early symptoms of chronic arsenic poisoning. In later stages, the disease can lead to gangrene, skin cancer, damage to vital organs, and eventually, death. In the months that followed, Chakraborti began to get letters from doctors and hospitals in Bangladesh, who were seeing streams of patients with similar symptoms. By 1995, it was clear that the country faced a crisis of untold proportions, and that the source of the poisoning was water from tubewells, from which 90 percent of the country gets its drinking water.

Experts estimate that today, arsenic in drinking water could threaten the health of 20 to 60 million Bangladeshis—up to half the country's population—and another 6 to 30 million people in West Bengal As many as 1 million wells in the region may be contaminated with the heavy metal at levels between 5 and 100 times the WHO drinking water guideline of 0.01 mg/liter.

How did the arsenic get into groundwater? Until the early 1970s, rivers and ponds supplied most of Bangladesh's drinking water. Concerned about the risks of water-borne disease, the WHO and international aid agencies launched a well-drilling program to tap groundwater instead. However, the agencies, not aware that soils of the Ganges aquifers are naturally rich in arsenic, didn't test the sediment before drilling tubewells. Because the effects of chronic arsenic poisoning can take up to 15 years to appear, the epidemic was not addressed until it was well under way.

Scientists are still debating what chemical reactions released the arsenic from the mineral matrix in which it is naturally bound up. Some theories implicate human activities. One hypothesis is that as water was pumped out of the wells, atmospheric oxygen entered the aquifer, oxidizing the iron pyrite sediments, and causing the arsenic to dissolve. An October 1999 article in the scientific journal *Nature* by geologists from the Indian Institute of Technology suggests that phosphates from fertilizer runoff and decaying organic matter may have played a role. The nutrient might have spurred the growth of soil microorganisms, which helped to loosen arsenic from sediments.

Salt is another naturally occurring groundwater pollutant that is introduced by human activity. Normally, water in coastal aquifers empties into the sea. But when too much water is pumped out of these aquifers, the process is reversed: seawater moves inland and enters the aquifer. Because of its high salt content, just 2 percent of seawater mixed with freshwater makes the water unusable for drinking or irrigation. And once salinized, a freshwater aquifer can remain contaminated for a very long time. Brackish aquifers often have to be abandoned because treatment can be very expensive.

In Manila, where water levels have fallen 50 to 80 meters [55 to 88 yards] because of overdraft, seawater has flowed as far as 5 kilometers [3 miles] into the Guadalupe aquifer that lies below the city. Saltwater has traveled several kilometers inland into aquifers beneath Jakarta and Madras, and in parts of the U.S. state of Florida. Saltwater intrusion is also a serious problem on islands such as the Maldives and Cyprus, which are very dependent on aquifers for water supply. Fluoride is another natural contaminant that threatens millions in parts of Asia. Aquifers in the drier regions of western India, northern China, and parts of Thailand and Sri Lanka are naturally rich in fluoride deposits. Fluoride is an essential nutrient for bone and dental health, but when consumed in high concentrations, it can lead to crippling damage to the neck and back, and to a range of dental problems. The WHO estimates that 70 million people in northern China, and 30 million in northwestern India are drinking water with high fluoride levels.

A CHEMICAL SOUP

With just over a million residents, Ludhiana is the largest city in Punjab, India's breadbasket state. It is also an important industrial town, known for its textile factories, electroplating industries, and metal foundries. Although the city is entirely dependent on groundwater, its wells are now so polluted with industrial and urban wastes that the water is no longer safe to drink. Samples show high levels of cyanide, cadmium, lead, and pesticides. "Ludhiana City's groundwater is just short of poison," laments a senior official at India's Central Ground Water Board.

Like Ludhiana's residents, more than a third of the planet's people live and work in densely settled cities, which occupy just 2 percent of the Earth's land area. With the labor force thus concentrated, factories and other centers of employment also group together around the same urban areas. Aquifers in these areas are beginning to mirror the increasing density and diversity of the human activity above them. Whereas the pollutants emanating from hog farms or copper mines may be quite predictable, the waste streams flowing into the water under cities contain a witch's brew of contaminants.

Ironically, a major factor in such contamination is that in most places people have learned to dispose of waste—to remove it from sight and smell—so effectively that it is easy to forget that the Earth is a closed ecological system in which nothing permanently disappears. The methods normally used to conceal garbage and other waste—landfills, septic tanks, and sewers—become the major conduits of chemical pollution of groundwater. In the United States, businesses drain almost 2 million kilograms [4.4 million lbs] of assorted chemicals into septic systems each year, contaminating the drinking water of 1.3 million people. In many parts of the developing world, factories still dump their liquid effluents onto the ground and wait for it to disappear.

In the Bolivian city of Santa Cruz, for example, a shallow aquifer that is the city's main water source has had to soak up the brew of sulfates, nitrates, and chlorides dumped over it. And even protected landfills can be a potent source of aquifer pollution: the EPA found that a quarter of the landfills in the U.S. state of Maine, for example, had contaminated groundwater.

In industrial countries, waste that is too hazardous to landfill is routinely buried in underground tanks. But as these caskets age, like gasoline tanks, they eventually spring leaks. In California's Silicon Valley, where electronics industries store assorted waste solvents in underground tanks, local groundwater authorities found that 85 percent of the tanks they inspected had leaks. Silicon Valley now has more Superfund sites—most of them affecting groundwater—than any other area its size in the country. And 60 percent of the United States' liquid hazardous waste—34 billion liters of solvents, heavy metals, and radioactive materials—is directly injected into the ground. Although the effluents are injected below the deepest source of drinking water, some of these wastes have entered aquifers used for water supplies in parts of Florida, Texas, Ohio, and Oklahoma.

Shenyang, China, and Jaipur, India, are among the scores of cities in the developing world that have had to seek out alternate supplies of water because their groundwater has become unusable. Santa Cruz has also struggled to find clean water. But as it has sunk deeper wells in pursuit of pure supplies, the effluent has traveled deeper into the aquifer to replace the water pumped out of it. In places where alternate supplies aren't easily available, utilities will have to resort to increasingly elaborate filtration set-ups to make the water safe for drinking. In heavily contaminated areas, hundreds of different filters may be necessary. At present, utilities in the U.S. Midwest spend $400 million each year to treat water for just one chemical—atrazine, the most commonly detected pesticide in U.S. groundwater. When chemicals are found in unpredictable mixtures, rather than discretely, providing safe water may become even more expensive.

ONE BODY, MANY WOUNDS

The various incidents of aquifer pollution described may seem isolated. A group of wells in northern China have nitrate problems; another lot in the United Kingdom are laced with benzene.

In each place it might seem that the problem is local and can be contained. But put them together, and you begin to see a bigger picture emerging. Perhaps most worrisome is that we've discovered as much damage as we have, despite the very limited monitoring and testing of underground water. And because of the time-lags involved—and given our high levels of chemical use and waste generation in recent decades—what's still to come may bring even more surprises.

Some of the greatest shocks may be felt in places where chemical use and disposal has climbed in the last few decades, and where the most basic measures to shield groundwater have not been taken. In India, for example, the Central Pollution Control Board (CPCB) surveyed 22 major industrial zones and found that groundwater in every one of them was unfit for drinking. When asked about these findings, CPCB chairman D. K. Biswas remarked, "The result is frightening, and it is my belief that we will get more shocks in the future."

Jack Barbash, an environmental chemist at the U.S. Geological Survey, points out that we may not need to wait for expensive tests to alert us to what to expect in our groundwater. "If you want to know what you're likely to find in aquifers near Shanghai or Calcutta, just look at what's used above ground," he says. "If you've been applying DDT to a field for 20 United

States years, for example, that's one of the chemicals you're likely to find in the underlying groundwater." The full consequences of today's chemical-dependent and waste-producing economies may not become apparent for another generation, but Barbash and other scientists are beginning to get a sense of just how serious those consequences are likely to be if present consumption and disposal practices continue.

CHANGING COURSE

Farmers in California's San Joaquin Valley began tapping the area's seemingly boundless groundwater store in the late-nineteenth century. By 1912, the aquifer was so depleted that the water table had fallen by as much as 400 feet [122 meters] in some places. But the farmers continued to tap the resource to keep up with demand for their produce. Over time, the dehydration of the aquifer caused its clay soil to shrink, and the ground began to sink—or as geologists put it, to "subside." In some parts of the valley, the ground has subsided as much as 29 feet [9 meters]—cracking foundations, canals, and aqueducts.

When the San Joaquin farmers could no longer pump enough groundwater to meet their irrigation demands, they began to bring in water from the northern part of the state via the California Aqueduct. The imported water seeped into the compacted aquifer, which was not able to hold all of the incoming flow. The water table then rose to an abnormally high level, dissolving salts and minerals in soils that had not been previously submerged. The salty groundwater, welling up from below, began to poison crop roots. In response, the farmers installed drains under irrigated fields—designed to capture the excess water and divert it to rivers and reservoirs in the valley so that it wouldn't evaporate and leave its salts in the soil.

But the farmers didn't realize that the rocks and soils of the region contained substantial amounts of the mineral selenium, which is toxic at high doses. Some of the selenium leached into the drainage water, which was routed to the region's wetlands. It wasn't until the mid-1980s that the aftermath of this solution became apparent: ecologists noticed that thousands of waterfowl in the nearby Kesterson Reservoir were dying of selenium poisoning.

Hydrological systems are not easy to outmaneuver, and the San Joaquin farmers' experience serves as a kind of cautionary tale. Each of their stopgap solutions temporarily took care of an immediate obstacle, but led to a longer-term problem more severe than the original one. "Human understanding has lagged one step behind the inflexible realities governing the aquifer system," observes USGS hydrologist Frank Chapelle.

Around the world, human responses to aquifer pollution thus far have essentially reenacted the San Joaquin Valley farmers' well-meaning but inadequate approach. In many places, various authorities and industries have fought back the contamination leak by leak, or chemical by chemical—only to find that the individual fixes simply don't add up. As we line landfills to reduce leakage, for instance, tons of pesticide may be running off nearby farms and into aquifers. As we mend holes in underground gas tanks, acid from mines may be seeping into groundwater.

Clearly, it's essential to control the damage we've already inflicted, and to protect communities and ecosystems from the poisoned fallout. But given what we already know—that damage done to aquifers is mostly irreversible, that it can take years before groundwater pollution reveals itself, that chemicals react synergistically, and often in unanticipated ways—it's now clear that a patchwork response isn't going to be effective. Given how much damage this pollution inflicts on public health, the environment, and the economy once it gets into the water, it's critical that emphasis be shifted from filtering out toxins to not using them in the first place. Andrew Skinner, who heads the International Association of Hydrogeologists, puts it this way: "Prevention is the only credible strategy."

To do this requires looking not just at individual factories, gas stations, cornfields, and dry cleaning plants, but at the whole social, industrial, and agricultural systems of which these businesses are a part.

The ecological untenability of these systems is what's really poisoning the world's water. It is the predominant system of high-input agriculture, for example, that not only shrinks biodiversity with its vast monocultures, but also overwhelms the land—and the underlying water—with its massive applications of agricultural chemicals. It's the system of car-dominated, geographically expanding cities that not only generates unsustainable amounts of climate-disrupting greenhouse gases and acid rain-causing air pollutants, but also overwhelms aquifers and soils with petrochemicals, heavy metals, and sewage. An adequate response will require a thorough overhaul of each of these systems.

Begin with industrial agriculture. Farm runoff is a leading cause of groundwater pollution in many parts of Europe, the United States, China, and India. Lessening its impact calls for adopting practices that sharply reduce this runoff—or, better still, that require far smaller inputs to begin with. In most places, current practices are excessively wasteful. In Colombia, for example, growers spray flowers with as much as 6,000 liters [1,585 gallons] of pesticide per hectare [a hectare is about 2.5 acres]. In Brazil, orchards get almost 10,000 liters [2,642 gallons] per hectare.

Experts at the U.N. Food and Agricultural Organization say that with modified application techniques, these chemicals could be applied at one-tenth those amounts and still be effective. But while using more efficient pesticide applications would constitute a major improvement, there is also the possibility of reorienting agriculture to use very little synthetic pesticide at all. Recent studies suggest that farms can maintain high yields while using little or no synthetic input. One decade-long investigation by the Rodale Institute in Pennsylvania, for example, compared traditional manure and legume-based cropping systems which used no synthetic fertilizer or pesticides, with a conventional, high-intensity system. All three fields were planted with maize and soybeans. The researchers found that the traditional systems retained more soil organic matter and nitrogen—indicators of soil fertility—and leached 60 percent less nitrate than the conventional system. Although organic fertilizer (like its synthetic

(continued on page 88)

Table 1: Some Major Threats to Groundwater

THREAT	SOURCES	HEALTH AND ECOSYSTEM EFFECTS AT HIGH CONCENTRATIONS	PRINCIPAL REGIONS AFFECTED
Pesticides	Runoff from farms, backyards, golf courses; landfill leaks.	Organochlorines linked to reproductive and endocrine damage in wildlife; organophosphates and carbamates linked to nervous system damage and cancers.	United States, Eastern Europe, China, India.
Nitrates	Fertilizer runoff; manure from livestock operations; septic systems.	Restricts amount of oxygen reaching brain, which can cause death in infants ("blue-baby syndrome"); linked to digestive tract cancers. Causes algal blooms and eutrophication in surface waters.	Midwestern and mid-Atlantic United States, North China Plain, Western Europe, Northern India.
Petro-chemicals	Underground petroleum storage tanks.	Benzene and other petrochemicals can be cancer-causing even at low exposure.	United States, United Kingdom, parts of former Soviet Union.
Chlorinated Solvents	Effluents from metals and plastics degreasing; fabric cleaning; electronics and aircraft manufacture.	Linked to reproductive disorders and some cancers.	Western United States, industrial zones in East Asia.

Threat	Sources	Health and Ecosystem Effects at High Concentrations	Principal Regions Affected
Arsenic	Naturally occurring; possibly exacerbated by over-pumping aquifers and by phosphorus from fertilizers.	Nervous system and liver damage; skin cancers.	Bangladesh, Eastern India, Nepal, Taiwan.
Other Heavy Metals	Mining waste and tailings; landfills; hazardous waste dumps.	Nervous system and kidney damage; metabolic disruption.	United States, Central America and northeastern South America, Eastern Europe.
Fluoride	Naturally occurring.	Dental problems; crippling spinal and bone damage.	Northern China, Western India; parts of Sri Lanka and Thailand.
Salts	Seawater intrusion; de-icing salt for roads.	Freshwater unusable for drinking or irrigation.	Coastal China and India, Gulf coasts of Mexico and Florida, Australia, Philippines.

Major sources: European Environmental Agency, USGS, British Geological Survey.

(continued from page 85)

counterpart) is typically a potent source of nitrate, the rotations of diverse legumes and grasses helped fix and retain nitrogen in the soil. Yields for the maize and soybean crops differed by less than 1 percent between the three cropping systems over the 10-year period.

In industrial settings, building "closed-loop" production and consumption systems can help slash the quantities of waste that factories and cities send to landfills, sewers, and dumps— thus protecting aquifers from leaking pollutants. In places as far-ranging as Tennessee, Fiji, Namibia, and Denmark, environ- mentally conscious investors have begun to build "industrial symbiosis" parks in which the unusable wastes from one firm become the input for another.

An industrial park in Kalundborg, Denmark diverts more than 1.3 million tons of effluent from landfills and septic systems each year, while preventing some 135,000 tons of carbon and sulfur from leaking into the atmosphere. Households, too, can become a part of this systemic change by reusing and repairing products. In a campaign organized by the Global Action Plan for the Earth, an international nongovernmental organization, thoughtful consumption habits have enabled some 60,000 households in the United States and Europe to reduce their waste by 42 percent and their water use by 25 percent.

As it becomes clearer to decisionmakers that the most serious threats to human security are no longer those of military attack but of pervasive environmental and social decline, experts worry about the difficulty of mustering sufficient political will to bring about the kinds of systemic—and there- fore revolutionary—changes in human life necessary to turn the tide in time. In confronting the now heavily documented assaults of climate change and biodiversity loss, leaders seem on one hand paralyzed by how bleak the big picture appears to be—and on the other hand too easily drawn into denial or delay by the seeming lack of immediate consequences of such delay. But protecting aquifers may provide a more immediate incentive for change, if only because it simply may not be possible to live with contaminated groundwater for as long as we could make do with a gradually more irritable climate or polluted air or impoverished wildlife. Although we've damaged portions of some aquifers to the point of no return, scientists believe that a large part of the resource still remains pure—for the moment. That's not likely to remain the case if we continue to depend on simply stepping up the present reactive tactics of cleaning up more of the chemical spills, replacing more of the leaking gasoline tanks, placing more plastic liners under landfills, or issuing more fines to careless hog farms and copper mines. To save the water in time requires the same fundamental restructuring of the global economy as does the stabilizing of the climate and biosphere as a whole—the rapid transition from a resource-depleting, oil- and coal-fueled, high-input industrial and agricultural economy to one that is based on renewable energy, compact cities, and a very light human footprint.

We've been slow to come to grips with this, but it may be our thirst that finally makes us act.

"HEAVEN IS UNDER OUR FEET"

Throughout human history, people have feared that the skies would be the source of great destruction. During the Cold War, industrial nations feared nuclear attack from above, and spent vast amounts of their wealth to avert it. Now some of that fear has shifted to the threats of atmospheric climate change: of increasing ultraviolet radiation through the ozone hole, and the rising intensity of global warming-driven hurricanes and typhoons. Yet, all the while, as the worldwide pollution of aquifers now reveals, we've been slowly poisoning ourselves from beneath. What lies under terra firma may, in fact, be of as much concern as what happens in the firmament above.

The ancient Greeks created an elaborate mythology about the Underworld, or Hades, which they described as a dismal, lifeless place completely lacking the abundant fertility of the world above. Science and human experience have taught us differently.

Hydrologists now know that healthy aquifers are essential to the life above ground—that they play a vital role not just in providing water to drink, but in replenishing rivers and wetlands and, through their ultimate effects on rainfall and climate, in nurturing the life of the land and air as well. But ironically, our neglectful actions now threaten to make the Greek myth a reality after all. To avert that threat now will require taking to heart what the hydrologists have found. As Henry David Thoreau observed a century-and-a-half ago, "Heaven is under our feet, as well as over our heads."

What Are the Threats to Your Drinking Water?

Does your drinking water come from a private well? If not, does your municipal water company draw water from a reservoir, a lake, or a river? Does your drinking water come from groundwater? Does your city have regulations to protect its drinking water supplies? Should it matter to you? If you like to drink clean water, it should.

"The EPA estimates that our nation needs $265 billion to maintain and improve its drinking water infrastructure over the next twenty years," said Senator James M. Jeffords. "If we don't address this, we'll be facing more and more health and environmental issues as our nation's water infrastructure degrades." The infrastructure for our water supplies refers to such things as the water treatment plants and the pipes through which water travels to get to the pipes in your house. There is another concern with drinking water supplies—the threat of contamination from multiple sources, including chemicals, pesticides, and human and animal wastes.

The following two articles explore the quality of our drinking water: The first article, "It's Your Drinking Water: Get to Know It and Protect It," explains the basics of how the Safe Water Drinking Act (SWDA) was passed in 1974 to protect our drinking water. The act authorized the U.S. Environmental Protection Agency (EPA) to set standards for protecting all wells that supply drinking water to more than 25 people. To safeguard our health, the EPA has set standards for safe levels of drinking water for more than 80 contaminants. Water suppliers are required to monitor and test the drinking water. The results of these tests are available for consumers to review in the water supplier's annual report. Since your body is about 70% water, it's worth investigating whether the water you are putting into it is clean.

The second article, "Is Your Drinking Water Safe?", offers some ideas to think about when you drink your next glass of tap water. The good news is that the United States can claim one of the best water supplies in the world. Even so, not everyone in the country has clean drinking water, and keeping the supplies clean is not easy. Recently in Washington, D.C., thousands of homes were found to have an unhealthy level of lead in their drinking water.

—The Editor

It's Your Drinking Water: Get to Know It and Protect It
from the U.S. Environmental Protection Agency

There is no such thing as naturally pure water. In nature, all water contains some impurities. As water flows in streams, sits in lakes, and filters through layers of soil and rock in the ground, it dissolves or absorbs the substances that it touches. Some of these substances are harmless. In fact, some people prefer mineral water precisely because minerals give it an appealing taste. However, at certain levels, minerals, just like man-made chemicals, are considered contaminants that can make water unpalatable or even unsafe.

Some contaminants come from erosion of natural rock formations. Other contaminants are substances discharged from factories, applied to farmlands, or used by consumers in their homes and yards. Sources of contaminants might be in your neighborhood or might be many miles away. Your local water quality report tells which contaminants are in your drinking water, the levels at which they were found, and the actual or likely source of each contaminant.

Some ground water systems have established wellhead protection programs to prevent substances from contaminating their wells. Similarly, some surface water systems protect the watershed around their reservoir to prevent contamination. Right now, states and water suppliers are working systematically to assess every source of drinking water and to identify potential sources of contaminants. This process will help communities to protect their drinking water supplies from contamination, and a summary of the results will be in future water quality reports.

WHERE DOES DRINKING WATER COME FROM?

A clean, constant supply of drinking water is essential to every community. People in large cities frequently drink water that comes from surface water sources, such as lakes, rivers, and reservoirs. Sometimes these sources are close to the community. Other times, drinking water suppliers get their water from sources many miles away. In either case, when you think about where your drinking water comes from, it's important to consider not just the part of the river or lake that you can see, but the entire watershed. The watershed is the land area over which water flows into the river, lake, or reservoir.

In rural areas, people are more likely to drink ground water that was pumped from a well. These wells tap into aquifers— the natural reservoirs under the earth's surface—that may be only a few miles wide, or may span the borders of many states. As with surface water, it is important to remember that activities many miles away from you may affect the quality of ground water.

Your annual drinking water quality report will tell you where your water supplier gets your water.

HOW IS DRINKING WATER TREATED?

When a water supplier takes untreated water from a river or reservoir, the water often contains dirt and tiny pieces of leaves and other organic matter, as well as trace amounts of certain contaminants. When it gets to the treatment plant, water suppliers often add chemicals called coagulants to the water. These act on the water as it flows very slowly through tanks so that the dirt and other contaminants form clumps that settle to the bottom. Usually, this water then flows through a filter for removal of the smallest contaminants like viruses and *Giardia*.

Most ground water is naturally filtered as it passes through layers of the earth into underground reservoirs known as aquifers. Water that suppliers pump from wells generally contains less organic material than surface water and may not need to go through any or all of the treatments described in the previous paragraph. The quality of the water will depend on local conditions.

The most common drinking water treatment, considered by many to be one of the most important scientific advances of the 20th century, is disinfection. Most water suppliers add chlorine or another disinfectant to kill bacteria and other germs.

Water suppliers use other treatments as needed, according to the quality of their source water. For example, systems whose water is contaminated with organic chemicals can treat their water with activated carbon, which adsorbs or attracts the chemicals dissolved in the water.

WHAT IF I HAVE SPECIAL HEALTH NEEDS?

People who have HIV/AIDS, are undergoing chemotherapy, take steroids, or for another reason have a weakened immune system may be more susceptible to microbial contaminants, including *Cryptosporidium*, in drinking water. If you or someone you know fall into one of these categories, talk to your health care provider to find out if you need to take special precautions, such as boiling your water.

Young children are particularly susceptible to the effects of high levels of certain contaminants, including nitrate and lead. To avoid exposure to lead, use water from the cold tap for making baby formula, drinking, and cooking, and let the water run for a minute or more if the water hasn't been turned on for six or more hours. If your water supplier alerts you that your water does not meet EPA's standard for nitrates and you have children less than six months old, consult your health care provider. You may want to find an alternate source of water that contains lower levels of nitrates for your child.

WHAT ARE THE HEALTH EFFECTS OF CONTAMINANTS IN DRINKING WATER?

EPA has set standards for more than 80 contaminants that may occur in drinking water and pose a risk to human health. EPA sets these standards to protect the health of everybody, including vulnerable groups like children.

The contaminants fall into two groups according to the health effects that they cause. Your local water supplier will alert you through the local media, direct mail, or other means if there is a potential acute or chronic health effect from compounds in the drinking water. You may want to contact them for additional information specific to your area.

Acute effects occur within hours or days of the time that a person consumes a contaminant. People can suffer acute health effects from almost any contaminant if they are exposed to extraordinarily high levels (as in the case of a spill). In drinking water, microbes, such as bacteria and viruses, are the contaminants with the greatest chance of reaching levels high enough to cause acute health effects. Most people's bodies can fight off these microbial contaminants the way they fight off germs, and these acute contaminants typically don't have permanent effects. Nonetheless, when high enough levels occur, they can make people ill, and can be dangerous or deadly for a person whose immune system is already weak due to HIV/AIDS, chemotherapy, steroid use, or another reason.

Chronic effects occur after people consume a contaminant at levels over EPA's safety standards for many years. The drinking water contaminants that can have chronic effects are chemicals (such as disinfection by-products, solvents, and pesticides), radionuclides (such as radium), and minerals (such as arsenic). Examples of these chronic effects include cancer, liver or kidney problems, or reproductive difficulties.

WHO IS RESPONSIBLE FOR DRINKING WATER QUALITY?

The Safe Drinking Water Act gives the Environmental Protection Agency (EPA) the responsibility for setting national drinking water standards that protect the health of the 250 million people who get their water from public water systems. Other people get their water from private wells which are not subject to federal regulations. Since 1974, EPA has set national standards for over 80 contaminants that may occur in drinking water.

While EPA and state governments set and enforce standards, local governments and private water suppliers have direct responsibility for the quality of the water that flows to your tap.

Water systems test and treat their water, maintain the distribution systems that deliver water to consumers, and report on their water quality to the state. States and EPA provide technical assistance to water suppliers and can take legal action against systems that fail to provide water that meets state and EPA standards.

WHAT IS A VIOLATION OF A DRINKING WATER STANDARD?

Drinking water suppliers are required to monitor and test their water many times, for many things, before sending it to consumers. These tests determine whether and how the water needs to be treated, as well as the effectiveness of the treatment process. If a water system consistently sends to consumers water that contains a contaminant at a level higher than EPA or state health standards or if the system fails to monitor for a contaminant, the system is violating regulations, and is subject to fines and other penalties.

When a water system violates a drinking water regulation, it must notify the people who drink its water about the violation, what it means, and how they should respond. In cases where the water presents an immediate health threat, such as when people need to boil water before drinking it, the system must use television, radio, and newspapers to get the word out as quickly as possible. Other notices may be sent by mail, or delivered with the water bill. Each water suppliers' annual water quality report must include a summary of all the violations that occurred during the previous year.

HOW CAN I HELP PROTECT DRINKING WATER?

Using the new information that is now available about drinking water, citizens can both be aware of the challenges of keeping drinking water safe and take an active role in protecting drinking water. There are lots of ways that individuals can get involved. Some people will help clean up the watershed that is the source of their community's water. Other people might get involved in wellhead protection activities to prevent the contamination of the ground water source that provides water to their community. These people will be able to make use of the information that states and water systems are gathering as they assess their sources of water.

Other people will want to attend public meetings to ensure that the community's need for safe drinking water is considered in making decisions about land use. You may wish to participate as your state and water system make funding decisions. And all consumers can do their part to conserve water and to dispose properly of household chemicals.

Is Your Drinking Water Safe?
by Rene Ebersole

It's clear, smells fresh, tastes good. But is it safe to drink? Most of us have considered this question, whether sipping tap water while traveling or making lemonade at home. Truth is American water supplies are some of the cleanest in the world. Still, it sure doesn't hurt to be cautious.

Case in point: The nation's capital, where thousands of residents discovered last winter [2003–2004] that their tap water was tainted with dangerous levels of lead, a metal associated with behavioral problems, brain damage and lowered IQ in children, and strokes, cancer and elevated blood pressure in adults.

More than two-thirds of the 6,000 Washington, D.C., households that had their water tested for lead were found to exceed the Environmental Protection Agency's (EPA) hazard level of 15 parts per billion (ppb)—157 homes had levels above 300 ppb. When the test results hit the news in January [2004], city officials scrambled to root out the source of the contamination and ramped up efforts to replace lead service pipes that transmit water underground to customers' homes.

At the same time, residents stormed department stores, looking for home water filtration systems that would protect their families from the invisible toxin.

The District of Columbia is home to one of the roughly three dozen water systems in the country with lead levels above the federal safety standard, according to data collected by the EPA. (Most of the others are in smaller communities.) But lead isn't the only contaminant in the nation's public water supplies.

Some studies estimate that as many as seven million Americans become sick from contaminated tap water each year. Often to blame are aging pipes that break, leach contaminants into water and breed bacteria, as well as old-fashioned treatment facilities that can fail to remove 21st century contaminants. Exacerbating the problem are environmental threats—stormwater runoff, agricultural pesticides and fertilizers, industrial pollution, hazardous waste and oil and chemical spills.

"Most Americans take it for granted that their tap water is pure and their water infrastructure is safe," says Erik Olson, senior attorney with the Natural Resources Defense Council (NRDC). "They shouldn't."

A recent NRDC report authored by Olson found the quality of drinking water in 19 major U.S. cities might pose health risks to some residents, especially those with compromised immune systems. Of all the cities studied, five had apparent or confirmed violations of enforceable tap water laws. The cities with the poorest water quality were Albuquerque, Boston, Fresno, Phoenix and San Francisco.

Water utilities are required by law to provide consumers with information about where municipal water comes from, whether it exceeds the allowable limits of the 80 possible contaminants regulated by the EPA and the health risks to which they may be exposed by drinking tap water. "If a system fails to monitor for a contaminant, it is subject to fines and penalties," says EPA spokesperson Cathy Milbourn. Utilities are required to mail these "consumer confidence reports" by July each year.

Can the utilities' reports be trusted? According to NRDC, sometimes yes, sometimes no. Olson and his colleagues carefully evaluated the utility reports of the 19 cities. They determined that some, such as Chicago, Denver and Detroit, were doing a "good job." Others, including Atlanta, Boston, Houston, Newark, Phoenix, Seattle and Washington, D.C., published information that was "incomplete or misleading."

Studies show that it will take from $230 to $500 billion—more than $300 million in Washington, D.C., alone—to restore and upgrade the nation's aging and outdated water systems. Some experts criticize the Bush administration for providing insufficient funding for such an important endeavor, and instead attempting to weaken the bedrock environmental laws that protect the nation's drinking water supplies.

At press time [July 2004], Congress was considering the administration's proposal to revoke Clean Water Act protections for more than 20 million acres [8.1 million hectares] of wetlands that naturally filter pollutants and provide clean drinking water to American towns and cities. The administration did, however, cancel plans last winter to issue a new rule limiting the types of waters protected under the Clean Water Act.

"Clean drinking water has been one of the major public health triumphs of the past 100 years," says David Ozonoff, a professor at Boston University's School of Public Health and an expert on waterborne illnesses. "We've figured out how to build very efficient water delivery systems. But these systems can either provide safe drinking water, or deliver poisons and harmful organisms into every home, school and workplace. One misstep can lead to disaster, so we must vigorously protect our watersheds and use the best technology to purify our tap water."

What Is Polluting Our Oceans?

We now have the scientific data to know that the things people do—from growing food and fertilizing their lawns, to disposing of waste and making plastics, to mining and burning coal—cause pollution in our oceans.

You might think that the major causes of pollution are oil tanker spills and garbage dumped out at sea. However, a report by the National Research Council (2002) showed that in one year, more oil flows off of our roads and parking lots (16 million gallons) than was dumped by the 1989 *Exxon Valdez* oil spill (10.9 million gallons). In fact, it is this nonpoint pollution that is the largest pollution source in our coastal waters. And the loss of our coastal wetlands, which act as biological filters, makes this pollution an even more serious problem. The loss of coastal wetlands results from a weakening of the protection that was once afforded coastal wetlands under the Clean Water Act.

The following chapter from the Pew Oceans Commission's 2003 report states that in 2001, more than 13,000 U.S. beaches were closed because of pollution or had pollution advisories. But the pollution is not just confined to the shores. Orca whales have high polychlorinated biphenyl (PCB) levels, and some species of deep-sea fish are too filled with mercury to be safe for people to eat. The Pew report states that today our oceans are facing as great a danger as our rivers and lakes did in the late 1960s and early 1970s, when some polluted rivers caught fire. Nitrogen run-off from fertilizers causes algae to bloom and allows seasonal "dead zones," or areas in which the fish have died from lack of oxygen, to appear. At times, the dead zone in the Gulf of Mexico is as large as the entire state of Massachusetts.

Point sources of pollution are still a problem. Regulations for cruise ships, ballast-discharge from ships, and concentrated animal feeding operations are sometimes poorly enforced. Invasive species and sound pollution also degrade our oceans.

But science tells us that we can improve. For example, coastal waters around Los Angeles and San Diego have dramatically improved over the last 25 years. Just as the creation of the Clean

Water Act in 1972 and its enforcement have improved freshwater quality, the Environmental Protection Agency (EPA) and the states must coordinate efforts to stem the flow of pollutants into our oceans.

The Pew Oceans Commission is an independent group of American leaders conducting a national dialogue on the policies needed to restore and protect living marine resources in U.S. waters. After reviewing the best scientific information available and speaking with people from around the country, the Commission made its formal recommendations to Congress in June 2003.

—The Editor

America's Living Oceans: Charting a Course for Sea Change

from the Pew Oceans Commission

THE NATURE OF THE PROBLEM

The images of the *Exxon Valdez* oil spill in Prince William Sound, Alaska, in 1989, and the sight of trash washing up with the seaweed on our favorite beaches are all too familiar.

What we are less aware of, however, is the amount of pollution that travels daily from each of our lawns, vehicle tailpipes, driveways, and the fields where our food is produced into our coastal waters. A recent [2002] study by the National Research Council found that the same amount of oil released in the *Exxon Valdez* spill—10.9 million gallons—washes off our coastal lands and into the surrounding waters every eight months. The Mississippi River, which drains nearly 40 percent of the continental United States, carries an estimated 1.5 million metric tons of nitrogen into the Gulf of Mexico each year. Overall, the amount of nitrogen released into coastal waters along the Atlantic seaboard and the Gulf of Mexico from anthropogenic, or human-induced sources, has increased about fivefold since the preindustrial era.

The consequences of this polluted runoff are most acute along the coasts, where more than 13,000 beaches were closed or under pollution advisories in 2001. Two-thirds of our estuaries and bays are either moderately or severely degraded from eutrophication. However, pollution's reach extends far beyond our major cities. Scientists report that killer whales have higher PCB levels in their blubber than any animal on the planet and that fish species that live their entire lives far out in the Pacific are too contaminated with mercury to be safe to eat.

These are the signs of a silent crisis in our oceans.

Fortunately, we have set a good precedent for addressing water pollution. In response to public outcry over such environmental calamities as the burning of the Cuyahoga River in Ohio, Congress passed the Clean Water Act (CWA) in 1972. The law requires the U.S. Environmental Protection Agency (EPA) to establish national technology standards and science-based criteria for water quality protection. The states then control identifiable sources of pollution by issuing pollution discharge permits based on these technology and water quality requirements.

Efforts resulting from the provisions of the Clean Water Act have succeeded in removing the worst pollution from the rivers and lakes that surround us. Some coastal waters, such as those off Los Angeles and San Diego, have dramatically improved. There, inputs of many pollutants have been reduced by 90 percent or more over a 25-year period, leading to the recovery of kelp beds, fish communities, and certain seabird populations.

But in the 30 years since the Clean Water Act was passed, as scientific knowledge and experience has improved, the focus of our concern has shifted. Although controlling point sources remains critical, the subtler problem of nonpoint sources has moved to the fore. In our oceans, now, we are experiencing a crisis as great as a burning river. It is a crisis we must address through changes in both policy and commitment.

Today, nonpoint sources present the greatest pollution threat to our oceans and coasts. Every acre of farmland and stretch of road in a watershed is a nonpoint source. Every treated lawn in America contributes toxins and nutrients to our coasts. Nonpoint pollutants include excess fertilizers and pesticides used in farming, oil and grease from paved surfaces, bacteria and nutrients from livestock manure, and acidic or toxic drainage from abandoned mines.

The current legal framework is ill equipped to address this threat. Rather than confronting individual cases, the situation requires that we apply new thinking about the connection between the land and the sea, and the role watersheds play in providing habitat and reducing pollution.

One of the major nonpoint pollutants is nitrogen, a nutrient that encourages plant growth. Although nitrogen is essential to life, in excess it can significantly damage and alter ecosystems. In fact, scientists now believe that nutrients are the primary pollution threat to living marine resources. Most nitrogen in the oceans arrives from nonpoint sources, including storm runoff from roads and agricultural fields, and airborne nitrogen emitted from power plants and car tailpipes.

We have also learned that marine species accumulate toxic substances. From single-celled marine life to top ocean predators, including humans, toxic substance levels in body tissue increase as predators consume contaminated prey. In addition, new forms of pollution are emerging. Non-native species, introduced by accident or design, have proliferated to stress entire ecosystems, crowding out native species, altering habitat, and in some instances, introducing disease. And human-generated sound in the oceans is affecting marine life in ways we are just beginning to understand.

Finally, we have not fully dispensed with the problem of point source pollution. Legal loopholes and poor enforcement allow significant point sources of pollution to go unregulated. These include cruise ships, ballast-water discharge from ships, and concentrated animal feeding operations. Animal feeding operations alone produce more than three times the amount of waste that people do—about 500 million tons of manure every year.

Through witness testimony from around the country, commissioned papers, and its own research, the [Pew Oceans] Commission investigated five types of pollution—nutrients, toxic substances, cruise ship discharges, invasive species, and anthropogenic sound. It reviewed the current state of our laws and changes necessary to control new and overlooked sources of pollution.

WHEN NUTRIENTS POLLUTE

The immediate cause of the 1991 event that killed one million menhaden in North Carolina's Neuse River was a single-celled creature called *Pfiestera piscicida*. Known as the killer alga, *P. piscicida* can emit a strong neurotoxin when in the presence of schools of fish. It feasts on the dead and dying fish, reproduces, and then settles back into the sediment. Scientists have found that *P. piscicida* thrives in coastal waters that are enriched with nutrients such as phosphorous and nitrogen.

The Neuse River outbreak was linked by analyses of the event to nutrients flowing from manure lagoons and other agricultural sources in the watershed.

We are degrading the environment along our coasts. Nutrient pollution has been linked to harmful algal blooms, such as the *Pfiestera* outbreak. It has also been linked to dead zones, such as the area in the Gulf of Mexico that appears annually and has reached the size of Massachusetts (more than 8,000 square miles [20,720 km^2]). In addition, this pollution results in the loss of seagrass and kelp beds, destruction of coral reefs, and lowered biodiversity in estuaries and coastal habitats. The incidence of harmful algal blooms along the United States coastlines increased from 200 in the decade of the 1970s to 700 in the 1990s, and now includes almost every coastal state in the U.S. One bloom off the coast of Florida was implicated in the deaths of more than 150 manatees.

The continued loss of wetlands is further evidence of this trend in degradation. Wetlands serve a critical function as natural filters that remove nutrients before they can reach the sea, but they are being lost at the rate of approximately 60,000 acres per year. If current practices of nutrient input and habitat destruction continue, nitrogen inputs to U.S. coastal waters in 2030 may be 30 percent higher than at present.

When too many nutrients—particularly nitrogen—enter the marine environment, the result is eutrophication—the overenrichment of the water that stimulates extraordinary growth of phytoplankton and attached algae. Phytoplankton blooms can be so dense they block the light needed by corals and by submerged vegetation such as seagrasses. Severe light deprivation will kill the plants and cause corals to expel the algae they host, which leads to coral bleaching.

After the phytoplankton die and sink to the ocean floor, bacteria decompose them. Decomposition pulls oxygen from the water, leaving the remaining plants and animals oxygen-starved. Areas with little oxygen, called hypoxic, are unable to support fish and shrimp populations, and the stress of hypoxia can make them more vulnerable to invasive species, disease, and mortality events. In addition to the well-known hypoxic dead zone at the mouth of the Mississippi River, hypoxic zones have developed in 39 estuaries around the U.S. coast.

Of the myriad sources of nutrient pollution, agriculture is the most significant. Nitrogen in fertilizer is easily dissolved in and transported by water. Animal wastes are also nitrogen rich, and are generally applied to farmland, where the nitrogen can be washed into water bodies by rainstorms. Aggravating this problem, tile drainage systems constructed to collect and shuttle excess water from fields—particularly common in the corn and soybean fields of the Midwest—provide an express-way for nitrogen flowing into waterways.

Until recently, atmospheric deposition—the settling of airborne pollutants on the land and water—has been an over-looked source of nitrogen pollution in coastal waters. It is now clear that it is widespread and quantitatively important in some regions. Most atmospheric deposition of nitrogen originates as nitrogen oxide emissions from power plants and automobiles, and ammonia gas released from animal wastes.

In addition to nonpoint sources, there are major point sources of nutrients, particularly concentrated animal feeding operations (CAFOs). Most animal wastes from CAFOs are stored in open lagoons, which can be larger than five and a half football fields and contain 20 to 45 million gallons of wastewater. If not properly managed, lagoons can leach nutrients and other substances into waterways and overflow during rainstorms. The liquid effluent, rich in nitrogen and phosphorous, is sprayed onto agricultural fields as fertilizer, often at many times the amount needed for crop growth. On a day-to-day basis, the over-application of animal waste to land, which fouls waterways with runoff, is a significant environmental problem.

Although they are regulated under the CWA [Clean Water Act], CAFOs have largely avoided pollution restrictions because of exemptions in outdated regulations and the states' failure to enforce permitting requirements. Of the approximately 15,500 operations that meet EPA's definition triggering regulation, less than 30 percent have permits, reducing the government's and the public's ability to monitor and control CAFO-related pollution. EPA recently revised its CAFO regulations, which now expressly require all CAFOs over a certain size to obtain a point source discharge permit. EPA's new regulations require CAFOs to develop a nutrient management plan by 2006, but EPA has not set enforceable standards for these plans, which will be written by the operators and not subject to government or public review. In exchange for developing and implementing a nutrient management plan, CAFOs are shielded from liability for pollution that is discharged off the facility's land application area.

Regardless of its source, nitrogen has become one of the most pervasive and harmful pollutants in coastal waters. A revitalized pollution policy must reflect this understanding.

TOXIC WATERS

When the *Exxon Valdez* ran aground in Alaska and spilled its oil cargo in March 1989, scientists, managers, and hundreds of volunteers rushed to rescue thousands of seabirds and sea otters. They picked the birds off soiled beaches and attempted to clean their plumage before the birds lost their ability to float and to stay warm. In the end, some 30,000 seabirds perished as well as 1,000 or more sea otters, and untold numbers of fish.

Congress has since passed the Oil Pollution Act to reduce the risk of similar tanker accidents.

New evidence strongly suggests that components of crude oil, called polycyclic aromatic hydrocarbons (PAHs), persist in the marine environment for years and are toxic to marine life at concentrations in the low parts-per-billion range. Chronic exposure to PAHs can affect development, increase susceptibility to disease, and jeopardize normal reproductive cycles in many marine species.

PAHs represent just one class of toxic substances that threaten the health of marine species and of humans who depend upon them for food. The Commission focused on three toxic substances of particular concern: PAHs, PCBs (polycholorinated biphenyls), and heavy metals like mercury. These substances are both pervasive and persistent. They are decomposed very slowly, if at all, by bacteria, and do not leave the marine environment quickly or completely. Although now banned in domestic manufacture of electrical transformers, plastics, paints, and other materials, PCBs are still present in many imported materials and at many industrial and military sites. Mercury levels are on the rise in some regions. Nearly 80 percent of the mercury in the marine environment arrives as air emissions from coal-fired power plants and other combustion sources, some of them overseas.

Landfills, urban runoff, ocean dumpsites, ocean vessels, and the burning of fossil fuels are just a few of the pathways that bring toxic substances to the oceans.

Toxic compounds enter marine food chains either directly from the water or from concentrated deposits in sediments. Organisms accumulate toxic substances in their tissues, where they may be passed up the food chain. Some of these compounds are concentrated at each step in the chain. The ocean's top predatory fish and marine mammals therefore often have the highest concentrations of toxic compounds in their bodies. Killer whales, walruses, and tuna are among those most contaminated.

Accumulated toxic substances disrupt hormone cycles, cause birth defects, suppress the immune system, and cause disorders resulting in cancer, tumors, and genetic abnormalities. In some instances, accumulated toxic substances can even cause death in marine animals.

The contamination of certain commercial species may pose particular problems for humans. Recent studies sponsored by *The Mobile Register* indicated that the presence of methyl-mercury (the bio-available form of mercury, and the form most prevalent in fish) in several species of fish in the Gulf of Mexico, including ling, amberjack, and redfish, may be so great that Food and Drug Administration standards would prohibit selling them to the public. In 2001, of the 2,618 fish advisories issued in U.S. waters, almost 75 percent were for mercury contamination. In Alaska and other polar regions, the evidence of correlation between increased toxic loads and declining health in humans and animals alike is mounting.

The Arctic and Antarctic are hard hit by certain persistent toxins, especially heavy metals and organochlorines, which include PCBs, due to the peculiar mechanisms by which these compounds are preferentially transported to the polar regions. Airborne toxins are repeatedly deposited and volatilized as they are swept by atmospheric circulation from their points of origin toward the polar regions. This process is known as the grasshopper effect because the substances "hop" from their sources to their ultimate repositories in the polar marine environment.

Not enough is being done to address the dangers that toxic substances pose to marine species and to humans. There are no water quality standards for PAHs under the CWA, no ambient air quality standards for mercury under the Clean Air Act (CAA), no systematic monitoring of toxins levels in most species consumed by humans, and there is insufficient effort to clean up toxic contaminants in sensitive marine environments. These policy shortcomings should be addressed without delay.

CRUISE SHIPS

Cruise ships can offer spectacular views and unparalleled wildlife experiences. For many Americans, cruises provide their only exposure to the oceans and marine wildlife, and the popularity of this activity is increasing. In recent years the cruise ship industry has grown at an average annual rate of eight percent, and expansion continues. In 2001, the North American cruise industry set a record when it carried 8.4 million passengers. In San Francisco Bay, a new cruise terminal is expected to more than double the number of ship visits per year. Cruise ships make frequent stops in Florida, the Caribbean, along the West Coast, Maine, and Alaska.

While taking a cruise can provide an invaluable experience for passengers, cruise ships can pose a particular risk to the very environments they seek to explore. With as many as 5,000 people onboard, a cruise ship is akin to a floating city, where people go about many of the same activities as they do at home: showering, cleaning, cooking. In addition, cruise ships offer such amenities as photo developing, hairdressing, and dry cleaning. The waste from these activities, however, is not regulated like waste produced from cities.

In one week, a typical cruise ship generates 210,000 gallons of black water (sewage), 1,000,000 gallons of gray water (shower, sink, dishwashing water), 37,000 gallons of oily bilge water, more than eight tons of solid waste, millions of gallons of ballast water containing potential invasive species, and toxic wastes from dry cleaning and photo processing laboratories. This effluent, when discharged untreated—as too often happens— delivers human pathogens, nutrients, and hazardous substances directly to the marine environment. The wastewater pollution from these ships is compounded by air pollution from burning trash and fuel emissions that enter the marine environment via atmospheric deposition.

Despite the fact that cruise ships discharge waste from a single source, they are exempted from regulation under the CWA point source permitting system.

The CWA allows the discharge of untreated black water anywhere beyond three miles from shore, and does not require any treatment of gray or ballast water. Only in Alaskan waters are cruise ships required to meet federal effluent standards;

treat gray water discharges; and monitor, record, and report discharges to state and federal authorities. In addition, the CWA authorizes the U.S. Coast Guard to inspect the discharge logs and pollution control equipment aboard ships. However, Coast Guard officers are not required to test discharges for compliance.

The CWA and the Act to Prevent Pollution from Ships together regulate bilge water, which must be run through an oil-water separator before it is discharged. The National Invasive Species Act encourages all oceangoing vessels to exchange ballast water but does not require them to do so. The air emissions from ships are covered under the CAA amendments of 1990, but the EPA has yet to impose regulations.

In short, the legal regime that covers cruise ships is complex but not comprehensive. Unless we take greater steps to control discharges and reduce pollution, we will continue to harm the very places we love to visit.

INVASIVE SPECIES

Invasive species—non-native species whose introduction harms or is likely to harm the environment, economy, or human health—present one of the most significant threats to biodiversity and healthy ecosystems. Once introduced, they have the potential to establish themselves alongside, or in place of, existing species. They can compete with native species for prey and habitat, facilitate the spread of diseases, introduce new genetic material, and even alter landscapes. Invasive species can impede endangered species conservation and restoration efforts. In the marine environment, some compete with commercially significant fish species for food and habitat, or they clog nets and eat bait. On land and in the sea, invasive species are responsible for about 137 billion dollars in lost revenue and management costs in the U.S. each year.

Invasive species are hard to identify and eradicate before they take hold in an ecosystem, which can occur remarkably quickly. For example, every 14 weeks, a new invasive species is discovered in the San Francisco Bay.

Ballast water is the primary vector for the release of invasive species into marine waters. Ballast water—and all the living creatures contained within it—is pumped into and out of oceangoing vessels for stabilization. Often it is taken up in one port and discharged in another. Every day, some 7,000 species are transported around the world via ballast water.

Another important vector is aquaculture. Species such as Atlantic salmon, grown on the western coasts of the U.S. and Canada, act as invasive species if they escape or are released unintentionally from aquaculture facilities into the surrounding waters. Once in the wild, they can compete with native species for food, shelter, and other resources, as well as spread disease. In some cases, species raised for aquaculture may interbreed with native species, potentially threatening the viability of native stocks.

Other vectors include the home aquarium industry, ship hulls, oil platforms, and marine debris. Invasive species arrive in seaweed used to pack live bait and via the pet trade industry. They also reach U.S. waters as live food imports. The Internet has significantly aided the introduction of new species. Today, consumers need only a credit card, access to a computer, and a delivery address to purchase marine life for food, for use as bait, or as pets.

In an increasingly global economy, all this mobility represents a serious threat to the health of living marine resources.

Our laws are not equipped to deal with these threats. Biological pollution by invasive species is the focus of the National Invasive Species Act of 1996 (NISA). However, under the NISA structure, invasive species are managed on a case-by-case, crisis-by-crisis basis, and the national focus is on terrestrial invasive species.

To the extent that NISA addresses marine species, it does so almost exclusively in the context of ballast-water discharges, despite the existence of many other vectors. Ballast-water exchange (BWE) is a procedure in which ships in the open ocean dump ballast water taken aboard in foreign ports. Its purpose is to lessen the chance of introducing coastal invasive species into potentially hospitable habitats in destination ports. However, BWE does not always dislodge species and it does not apply to coastwise travel, which can also allow species to be transported to new environments. Additionally, BWE is not mandatory under NISA. Although the U.S. Coast Guard is required to check ship logs to determine whether an exchange occurred, it is not required to check the ballast tanks. Current guidelines encourage ship operators to report voluntary exchange, but compliance with this minimal requirement is weak.

There is little law focusing on other vectors of invasive species. For example, there is no uniform regime in place to track live imports either entering or traveling around the country. There is no systematic process for determining which management approach is best when a species is found, no central source of information for researching species, and no dedicated source of funding to control invasive species. For species like the destructive seaweed, *Caulerpa taxifolia*, which grows as much as three inches [7.6 cm] a day, any delay in response could have severe environmental and economic ramifications.

Currently, agencies at different levels of government report commodities using a different nomenclature and verification system. With such inconsistency, neighboring states could simultaneously be working to promote and eradicate the same species, and one agency's food list could be another agency's most wanted list of invaders. The lack of regulatory clarity was brought home by the discovery of the invasive snakehead fish in a Maryland pond. Federal regulations did not prohibit the importation or interstate transportation of this Asian fish and state law provided only a mild penalty for release of the fish, for which the statute of limitations had expired. Furthermore, state managers had no clear legal authority to eradicate the population that had established itself. This type of confusion results in invasive species—literally—slipping through the regulatory cracks and getting into the environment without anyone noticing.

SOUNDS

The use of anthropogenic sound as a tool in the ocean has become enormously valuable for scientists, engineers, fishermen, and the military. It allows fishermen to locate schools of fish and to keep predators from raiding or becoming entangled in their nets. The use of sound also helps mariners detect icebergs and other obstructions, biologists study behavior changes in marine species, oceanographers map the bottom of the ocean floor, geologists find oil and gas, climatologists research global climate change, and the U.S. Navy detect submarines.

Many marine species, including marine mammals, turtles, and fish, also rely on sound. They use vocalizations and their ability to hear to detect predators, prey, and each other. In the oceans, as on land, sound is essential for communication.

Anthropogenic sound in the ocean is on the rise, mainly due to increased vessel traffic. Coastal development is bringing more pleasure craft, and globalization and international trade require more commercial vessels. In addition, the navies of the United States and other nations are increasingly using active sonar systems to patrol coastal waters for enemy submarines. Meanwhile, oil and gas operations on the outer continental shelf are expected to spread into deeper waters. Climate change, too, may have a significant effect on sound levels in the ocean. Not only does sound travel faster in warmer water, but also rising temperatures and melting ice at the poles may open new shipping channels in areas that have previously experienced little vessel traffic.

Sound sources differ in both their intensity and frequency, and thus can have varied effects on species. Sounds in the same frequency ranges used by marine species can mask acoustic communication among animals and interfere with detection of prey and predators. High-intensity sounds can cause pain and, in some circumstances, tissue and organ damage. If the pressure resulting from the sound is intense enough, the animal can experience internal bleeding and subsequent death.

A mass stranding of whales in 2000 heightened concerns about the effects of sound in the oceans. In March of that year, at least 17 whales were stranded on beaches in the northern Bahama Islands. Most of the animals were alive when they stranded and eight of them were returned to the sea. The other nine animals died; pathology reports revealed bruising and internal organ damage. The stranding occurred about the time that ten U.S. Navy vessels were operating their midfrequency sonar systems nearby. Investigations conducted cooperatively by the Navy and the National Marine Fisheries Service suggested that the sonar transmissions were a critical factor in the strandings.

Low-intensity sounds can disrupt behavior and cause hearing loss, ultimately affecting longevity, growth, and reproduction. Frequent or chronic exposure to both high- and low-intensity sounds may cause stress, which human and terrestrial animal studies indicate can affect growth, reproduction, and ability to resist disease. Impulse sounds, such as those produced by explosions and seismic air guns, may damage or destroy plankton, including fish eggs and larvae, as well as damage or destroy tissues and organs in higher vertebrates.

The Marine Mammal Protection Act (MMPA), Endangered Species Act (ESA), and the National Environmental Policy Act (NEPA) all provide legal mechanisms for addressing sound. However, the MMPA and ESA apply only to marine mammals and endangered species, and are only capable of protecting individuals from particular sound-related projects, such as drilling operations or sonar activities. In addition, the federal government has recently proposed to exempt certain activities from environmental review under NEPA. Because review under these statutes is triggered only on a case-by-case basis and does not effectively address cumulative impacts on marine ecosystems, underwater sound as a source of potentially significant pollution in the marine environment has not received comprehensive treatment. A new policy framework is needed to adequately address this emerging pollution concern.

ACTION TO REDUCE MARINE POLLUTION

For too long our oceans have been dumping grounds. Within U.S. waters, ecosystems are subjected to insults from nonpoint, unregulated point, and nontraditional types of pollution from both land- and ocean-based sources. Nutrients, toxins, cruise ship discharges, acoustic and biological pollution, and invasive species all harm marine ecosystems, and the legal regimes in place do not match the nature of today's pollution threats. For each of these pollution sources, policy changes can and should be made as quickly as possible.

SUMMARY OF RECOMMENDATIONS

1. Revise, strengthen, and redirect pollution laws to focus on nonpoint source pollution on a watershed basis.

EPA and the states should establish water quality standards for nutrients, especially nitrogen, as quickly as possible. EPA and the states should also ensure that water quality standards are in place for other pollutants—such as PAHs, PCBs, and heavy metals such as mercury—where these are identified as problematic on a watershed-by-watershed basis. Congress should amend the Clean Water Act to require the use of best management practices to control polluted runoff resulting from agriculture and development. Congress and the executive branch should provide substantial financial and technical support for the adoption of such practices. Congress should link the receipt of agricultural and other federal subsidies to compliance with the Clean Water Act. Finally, Congress and the Environmental Protection Agency should ensure that air emissions of nitrogen compounds, mercury, and other pollutants are reduced to levels that will result in a substantial reduction of their impact on marine ecosystems.

2. Address unabated point sources of pollution.

Concentrated animal feeding operations should be brought into compliance with existing provisions in the CWA. Congress should enact legislation that regulates wastewater discharges from cruise ships under the CWA by establishing uniform minimum standards for discharges in all state waters and prohibiting discharges within the U.S. Exclusive Economic Zone that do not meet effluent standards. Congress should amend

NISA to require ballast-water treatment for all vessels that travel in U.S. waters, and regulate ballast-water discharge through a permitting system under the CWA. Finally, the International Maritime Organization draft convention on ballast-water management should be finalized and its provisions implemented through appropriate U.S. laws.

3. Create a flexible framework to address emerging and nontraditional sources of pollution.

A national electronic permitting system should be created under NISA to facilitate communication and track imports of live species that may result in aquatic introductions. Each state should inventory existing species and their historical abundance, in conjunction with the development of the regional ocean governance plans under the National Ocean Policy Act. Congress should provide adequate funding for developing statewide invasive species management plans that include provisions for inventorying, monitoring, and rapid response. With regard to sound, a comprehensive research and monitoring program should be developed to determine the effects of sound sources on living marine resources and ecosystems. Consideration should be given to requiring the utilization of best available control technologies, where the generation of sound has potential adverse effects. Finally, the environmental ramifications of any sound-producing project should be taken into formal consideration—pursuant to NEPA or other applicable statutes—at the planning stages of the project, before significant resources, time, and money have been devoted to its development.

4. Strengthen control of toxic pollutants.

The U.S. should ratify the Stockholm Convention on Persistent Organic Pollutants (POPs), and implement federal legislation that allows for additions to the list of the "dirty dozen" chemicals. In concert with this effort, EPA should develop and lead a comprehensive monitoring program to quantify levels of particular toxic substances in designated ocean habitats and species, and sufficient resources should be devoted to studying the effects of toxins on marine species. This monitoring program should be coordinated with Food and Drug Administration [FDA] and EPA seafood contaminant advisory efforts, so that people know where their seafood comes from and what it contains.

How Clean Are Our Nation's Beaches?

Usually, if you want to go swimming, you probably check the weather report, and not the pollution advisory, for the beach. But things have changed. According to a report released in August 2004 by the Natural Resources Defense Council (NRDC), more beach closings and advisory days were posted in 2003 than in any of the previous 14 years that NRDC had been monitoring them. The beaches included in the report are in U.S. coastal areas and on the Great Lakes. The numbers show a 51% increase in closures over 2002. Part of the increase is because of better pollution monitoring programs, but there is no escaping the fact that beaches are under serious threat.

The Natural Resources Defense Council is a national, nonprofit organization of scientists, lawyers, and environmental specialists dedicated to protecting public health and the environment. It was founded in 1970. The NRDC press release for its 2004 report noted the disturbing finding that local authorities could not identify the pollution sources that caused 68% of the beach closings and pollution advisories. This was the highest rate of "unknown sources" in the 14 years that NRDC has done this survey.

The report echoed what other reports have found: Pollution from the runoff of paved surfaces, such as parking lots, roads, and rooftops, is the "largest and fastest growing source of water pollution in coastal waters." In addition to nonpoint pollution sources, sewage overflows and malfunctioning sewage treatment plants and pump stations pollute coastal waters. Septic systems, boating waste, oil pollution, and waterfowl are other contributing sources.

There are actions that can be taken on the local, state, and federal levels to address this problem. The public outcry in the early 1970s over filthy lakes and rivers encouraged Congress to pass the Clean Water Act in 1972. Thanks to this historic legislation, there already exists a legal toolbox to begin curbing sources of water pollution. Now it is time to put it to use to take better care of our ocean resources.

—The Editor

Testing the Waters 2004: A Guide to Water Quality at Vacation Beaches

from the Natural Resources Defense Council

Our nation is blessed with nearly 23,000 miles [37,014 km] of ocean shoreline (excluding Alaska), more than 5,500 miles [8,851 km] of Great Lakes shoreline, and 300,000 miles [482,803 km] of rivers (many of which eventually empty into oceans and bays). Beaches are one of America's favorite vacation destinations. More than 180 million Americans visit coastal and Great Lakes counties and their beaches each year, generating tens of billions of dollars in sales of goods and services and supporting tens of millions of jobs. According to the April 2004 Preliminary Report by the U.S. Commission on Ocean Policy, "[coastal] tourism and recreation constitute some of the fastest growing business sectors—enriching economies and supporting jobs in communities virtually every-where along the coasts of the continental United States, southeast Alaska, Hawaii, and our island territories and commonwealths."

While America's waters are cleaner now than they were 30 years ago when rivers were burning and lakes were dying, U.S. coastal waters are generally in fair to poor condition. National water-quality monitoring data reported by the states in the year 2000, the most recent national picture available from the Environmental Protection Agency (EPA), show that approximately 45 percent of waters assessed by the states are not clean enough to support basic uses such as fishing or swimming; i.e., they do not meet water quality standards. To make matters worse, our towns and our states do not necessarily protect us from swimming-associated diseases, such as gastroenteritis, hepatitis, salmonellosis, and other infections and viruses.

America's beachgoers should feel confident that a visit to the beach will not be followed by a visit to the doctor. We need to know that the waters in which we swim, surf, and dive are safe. At a minimum, this means that recreational waters must be regularly tested, and the results must be measured against effective health standards. When waters do not meet these standards, the public needs to be promptly and clearly notified.

To spotlight this issue, every summer for the past 14 years the Natural Resources Defense Council (NRDC) has undertaken a study of beach closings/advisories and beachwater monitoring and public-notification programs in coastal and Great Lakes states. National and local policymakers, reporters, and activists have come to rely on NRDC's *Testing the Waters* reports for reliable, up-to-date information. Our work has triggered significant expansion of beachwater monitoring programs across the United States and new and emerging laws—particularly the federal BEACH Act of 2000, which is better protecting the health of beachgoers. At least 12 states initiated or expanded monitoring programs between the time NRDC began the report in 1991 and the passage of the BEACH [Beaches Environmental Assessment and Coastal Health] Act (Alabama, California, Florida, Georgia, Iowa, Maine, Massachusetts, Mississippi, North Carolina, Ohio, South Carolina, and Texas). As a result of federal grants now available to states through the BEACH Act, virtually every coastal and Great Lakes state is in the process of either initiating or expanding monitoring and public notification programs.

While beaches are better monitored than in the past, that monitoring reveals the extent to which sources of beachwater pollution remain unaddressed. Closings and advisories continue to rise while the sources of that pollution are usually not even identified, much less controlled. NRDC is issuing this report—the 14th annual *Testing the Waters*—as a reminder of how unchecked pollution continues to lower Americans' quality of life and threatens public health. This year [2004], as in past years, we found that water pollution continues to degrade the quality and health of many of our nation's ocean, bay, Great Lakes, and other freshwater beaches. Too often, rainwater turns to ruinous water as it washes untreated sewage and other contaminants into our oceans, bays, and lakes. Significant stretches of coastline are subject to closings or swimming advisories during the year. We hope this report spurs Congress, the Environmental Protection Agency (EPA), states, and localities to improve prevention and control over the sources of coastal pollution and to close remaining gaps in monitoring and public notification at our nation's beaches.

CHAPTER 1
SOURCES OF BEACHWATER POLLUTION

Most beach closings and advisories are based on monitoring that detects elevated levels of bacteria indicating the presence of microscopic disease-causing organisms from human and animal wastes. These wastes typically enter coastal waters from discharges of untreated or partially treated wastes from sewage treatment plants and sanitary sewers (such as combined sewer or sanitary sewer overflows), septic system failures, and stormwater runoff from urban, suburban, and rural areas. Beach closings and advisories can also be responses to other specific pollution events, such as an overflow from an animal-waste lagoon or an oil spill. In some coastal and Great Lakes counties, advisories are issued as a precautionary measure when storms occur because past experience demonstrates that rain carries pollution to swimming waters.

Eighty-eight percent of beach closings/advisories in 2003 were based on monitoring that detected bacteria levels exceeding beachwater quality standards. Six percent were precautionary due to rain that was known to carry pollution into coastal waters, 4 percent of beach closings and advisories were issued in response to a known pollution event (without relying solely on monitoring results), and 2 percent were due to other sources, such as dredging and algal blooms. The percentages may add up to more than 100 percent because more than one cause may have contributed to a closing or advisory. For example, some closings/advisories begin as precautionary measures due to rain or a known contamination event, and are then extended because subsequent monitoring reveals prolonged elevated bacteria levels.

The major pollution sources, as well as the number of closings and advisories in 2003 attributable to these sources, are summarized in Table 1.

Sewage overflows from sanitary and combined sewers and malfunctioning sewage treatment plants and pump stations have always been a major cause of ocean, bay, and Great Lakes beach closings and advisories. Today, stormwater runoff from

urban, suburban, and nonurban areas is also recognized as a
significant problem, resulting in elevated bacteria levels and
increased illness rates for swimmers. Almost every coastal and
Great Lakes state reported having at least one beach where
stormwater drains onto or near bathing beaches. New Jersey,
California, Florida, and Connecticut are among the states that
reported having numerous beaches near stormwater outfalls.

Trends in population and development indicate that by
2010, more than half of the people in the United States will live in

Table 1: Major Pollution Sources Causing Beach Closings/Advisories in 2003

POLLUTION SOURCE	NUMBER OF CLOSINGS/ADVISORIES*
Elevated bacteria levels of unknown origin	16,120 days plus 62 extended and 53 permanent events
Stormwater and runoff	2,616 days plus 2 extended and 17 permanent events
Sewage spills and overflows[†]	1,820 days plus 6 permanent events
Other (algal blooms dredging, wildlife, etc.)	268 days plus 1 extended and 1 permanent event
Rain or preemptive[††]	1,142 days

[*] Total exceeds national total of 18,284 because more than one source may apply to a given closing/advisory.

[†] Includes sewage overflows from combined and sanitary sewers, malfunctioning sewage treatment plants and pump stations, sewage spills, and sewer-line breaks.

[††] Usually due to stormwater or sewer overflows.

Extended: closings of 7 to 13 weeks. Permanent: closings of more than 13 weeks.

coastal towns and cities. Coastal population grew by 37 million people between 1970 and 2000 and by 2015 is projected to increase by another 21 million. As population grows along the U.S. coast, sewage systems become more stressed, and impervious surfaces—roofs, roads, and parking lots, for example—increase. Unless strong measures are taken, sewer overflows and stormwater runoff from these rapidly growing areas will increasingly degrade coastal waters and pollute our beaches.

The following is a description of the different sources of beachwater pollution.

COMBINED SEWER OVERFLOWS

Combined sewers are pipes designed to carry both raw sewage from residences and industrial sites and stormwater runoff from streets to sewage treatment plants. When it rains—even as little as one-quarter of an inch [just over half a centimeter]—the volume of the combined wastewater becomes too great for the treatment plant to handle. In such circumstances, the flow is then diverted to outfall points that discharge pollutants—including raw sewage; floatables such as garbage, syringes, and tampon applicators; toxic industrial waste; and contaminated stormwater—into the nearest stream or coastal waterway. These untreated discharges can be as potent as direct sewer emissions.

Combined sewer overflows (CSOs) are one of the major causes of pathogen contamination in marine and Great Lakes waters near urban areas. Although they are most prevalent in urban areas, CSOs occur throughout coastal and Great Lakes states. Combined sewer overflows also contaminate shellfish waters and recreational beaches. Combined sewers serve 40 million people nationwide. Although an EPA policy that aims to reduce these overflows has been in effect since 1994, virtually all combined sewer systems continue to overflow when it rains. The EPA 1994 Combined Sewer Overflow Control policy required all systems to implement nine minimum system controls by 1997. According to the EPA's 2001 Report to Congress, progress has been slow. Only 32 percent of CSO communities are implementing the nine minimum system controls, despite the January 1997 deadline. Only 19 percent have finished their plans for controlling CSOs, and fewer than 10 percent have finished implementing CSO controls. The EPA estimates that 1,260 billion gallons of raw sewage from CSO discharges continue to flow into our surface waters every year.

SANITARY SEWER OVERFLOWS AND SEWER-LINE BREAK DISCHARGES

Many sanitary sewers—those designed to carry only human and industrial waste from buildings to sewage-treatment plants—pose a threat to bathing-beach safety. Human-waste sewage lines have breaches, obstructions such as tree roots or grease clogs, cracks, stormwater drain cross-connections, and open manholes that permit infiltration by groundwater and inflows of stormwater. These sanitary sewers can become overloaded, especially when it rains, and can overflow and discharge raw sewage from manholes, manhole bypasses, pump-station bypasses, and treatment-plant bypasses. In addition, sanitary-sewer lines are often old and, in many cities, inadequately maintained. They can break and spill sewage directly onto streets or into waterways. The EPA has estimated 40,000 sanitary sewer overflows (SSO) annually and about 400,000 basement backups. These overflows often discharge sewage, untreated, directly into coastal waterways or their tributaries, causing between 1.8 and 3.5 million people to get gastroenteritis each year from swimming in raw sewage from SSOs. In January 2001, the EPA proposed SSO regulations that would require improved capacity, operation, and maintenance and require that systems notify the public when these overflows occur. Unfortunately, instead of finalizing rules to prevent SSOs, the Bush Administration has held up these proposed rules.

To make matters worse, the EPA has recently proposed to allow the "blending" of minimally treated sewage with secondarily treated effluent during rains. In effect this policy would allow the discharge of nearly raw sewage so long as it is sufficiently diluted. Analyses of pathogen data from blended sewage discharges gathered by the Milwaukee Public Health Department conclude that people swimming at the sewage discharge point in Lake Michigan would have had a 50 percent chance of contracting giardiasis after this blended sewage had been released, representing a thousandfold increase in the risk of getting sick as a result of exposure to pathogens while swimming.

SEWAGE TREATMENT PLANT MALFUNCTIONS

Sewage plants in coastal areas tend to serve densely populated, rapidly growing urban areas. When too many homes and businesses are hooked up to a sewage treatment plant, the plant cannot treat the sewage adequately. The National Oceanic and Atmospheric Administration (NOAA) estimates that between 1990 and 2010 the coastal population will grow from 112 million to more than 127 million—an increase of almost 13 percent. Plants that exceed their capacity or the capacity of their collection systems are prone to more frequent episodes of bypasses and inadequate treatment. Moreover, sewage treatment plants can, and often do, malfunction as the result of human error, breakage of old equipment, or unusual conditions in the raw sewage. At such times, raw or partially treated sewage may be discharged into coastal waterways and their tributaries. Sewage plants are increasingly discharging inadequately treated sewage in high-flow conditions. Instead of providing full secondary treatment for sewage, treated and untreated sewage are mixed, increasing the pathogen load to receiving waters.

URBAN STORMWATER RUNOFF

Stormwater starts as rain or snowmelt. As it washes over roads, rooftops, parking lots, construction sites, and industrial or commercial sites, it becomes contaminated with oil and grease, heavy metals, pesticides, litter, and pollutants from vehicle exhaust. On its way to stormdrains, it also often picks up fecal matter from dogs, cats, pigeons, other urban animals, and even humans. Stormdrains may empty into separate storm sewer systems that carry only stormwater and discharge directly into waterways, or they may become a part of combined sewer systems, which, as described above, usually overflow when it rains. Moreover, human waste may find its way into storm sewer systems from adjacent sewage pipes that leak or from businesses or residences that are illegally hooked up. Illicit discharges may also include such substances as oil from cars, paint, and grease from restaurants. In Los Angeles County, for instance, the sewer system is separate from the stormdrain system, yet Santa Monica Bay stormdrains sometimes discharge runoff containing human enteric viruses, indicating the presence of human wastes. About a quarter of our nation's polluted estuaries and lakes are fouled by urban stormwater, and it is a significant source of bathing-beach pollution in many regions. Almost every coastal or Great Lakes state has beaches with stormdrains nearby.

Urban runoff can cause swimmers to become ill. A recent study of the health effects of urban runoff compared illness rates for surfers in California's urban North Orange County and rural Santa Cruz County during the winters of 1998 and 1999. The study determined that surfers at the North Orange County beaches, which were more polluted by urban runoff, saw modestly elevated incidences of illness in 1999, and twice as many illnesses in 1998 (likely because 1998 was a wet year associated with El Niño). The EPA regulations require cities and industrial and construction sites to obtain permits, develop stormwater management plans, and implement best management practices; however, only limited progress has been made to date. Vigorous implementation and enforcement and ambitious pollutant reduction goals are necessary to make this program successful. Unfortunately, despite the magnitude of stormwater pollution, the EPA recently declined to set baseline technology standards for new construction and development.

POLLUTED RUNOFF FROM NONURBAN AREAS

In nonurban and suburban areas, rainwater often flows directly over farms, roads, golf courses, and lawns into waterways.

Farm runoff may contain high concentrations of pathogenic animal waste, fertilizers, and pesticides. Suburban lawn runoff often contains significant amounts of animal waste, fertilizer, and other chemicals. This uncontrolled runoff can foul beaches in less densely populated areas. Animal waste can also contain pathogens usually not found in human waste, such as *E. coli* 0157:H7. Animal waste from large feedlots, often spread too heavily on fields, runs off and has been linked to outbreaks of a toxic microorganism, *Pfiesteria piscicida*, in the Chesapeake Bay region and North Carolina, causing numerous waterway closings.

SEPTIC SYSTEMS

Dwellings built near the coast may be equipped with underground septic systems, which, if not sited, built, and maintained properly, can leach wastewater into coastal recreational waters. Homeowners often do not adequately maintain their septic systems, and there is no federal regulatory program to control waste from septic systems. Local governments and states rarely inspect septic systems sufficiently to prevent such failures. Bathing beaches can be contaminated by fecal matter from malfunctioning or overloaded septic systems. Runoff can also carry bacteria from failing septic systems far from the shore into streams that empty into bays near beaches. The EPA estimates that 25 percent of all U.S. dwellings use some kind of septic tank or on-site waste disposal system.

BOATING WASTES

Improperly handled boating wastes can pose a health and aesthetic threat to bathing beaches. Elevated concentrations of fecal coliform have been found in areas with high boating density. Federal law requires boats with onboard toilets either to treat the waste (through chemical treatment) before discharging it or to hold the waste and later pump it out for treatment at a sewage-treatment plant. Many areas lack sufficient pump-out facilities, and compliance with the law appears to be poor in many areas.

OIL POLLUTION

Oil reaches the ocean as a result of the production, transportation, and use of oil, and even from natural seeps. Oil spilled during tanker accidents, pipeline breaks, and refinery accidents can foul beaches. Many oils evaporate quickly, creating unsafe fumes. Other oils form globules that can float for days and wash onto beaches for weeks after a spill. In addition to the catastrophic oil spills, there are small oil spills every time it rains. Over the course of a year, urban runoff from a city of 5 million [people] can contain as much oil and grease as a large tanker spill. Normal ship operations also frequently cause small oil spills. A recent report by the National Research Council states that nearly 85 percent of the 29 million gallons of petroleum that enter North American ocean waters each year as a result of human activities comes from land-based runoff, polluted rivers, airplanes, and small boats and jet skis. The report estimated that two-stroke outboard motors release between 0.6 and 2.5 million gallons of oil and gasoline into U.S. coastal waters each year. Oil runoff from cars and trucks is increasing in coastal areas where the population is increasing and roads and parking lots are expanding to accommodate this growth.

WATERFOWL

Several municipalities listed waterfowl as the cause of a beach closing or advisory. Large or excessive populations of waterfowl at a beach or in suburban areas draining near a beach can occur during migration season when other potential waterfowl habitats are unavailable for some reason, such as filling of wetlands, or at other times as a result of altered ecological conditions (for example, when Canada geese that were previously migratory become resident). The fecal matter from these waterfowl can sometimes overload the normal capacity of a beach to absorb natural wastes, degrading water quality.

How Can We Reduce Oil Pollution in Our Oceans?

Welcome news is that there is less oil in the sea now than there was in 1985, according to the National Research Council (NRC). Even more compelling is evidence that the largest source of oil in the oceans is not from oil spills, but from nonpoint pollution—what washes off the roads and other paved surfaces, into the storm drains, and into the oceans. This accounts for about 16 million gallons of oil a year washing into U.S. coastal waters.

No matter how it gets into the ocean, oil has been an environmental problem for a long time. Large oil tanker spills, like the 10.9 million gallon spill from the ship *Exxon Valdez* in 1989, kill hundreds of thousands of animals, including fish, birds, and mammals, and ruin thousands of miles of coastline. But the toxic oil, and the toxic compounds created as the oil degrades over time, can remain in the water and sediment for many years. For example, oil from a barge spill 30 years ago is still present just inches below the sediment in some Massachusetts salt marshes. This is a concern because even small amounts of oil can harm marine organisms.

The following article was written by Nancy Rabalais, a professor at the Louisiana Universities Marine Consortium. Rabalais chairs the National Research Council's Ocean Studies Board. The National Research Council is part of The National Academies, established in 1916 as a private, nonprofit institution that provides science, technology, and health policy advice under a congressional charter.

As Rabalais's article notes, the threat of oil in the oceans must be minimized by reducing spills and by decreasing the inputs from nonpoint sources. The author cites the need for regulatory and safety measures to address the potential spills from aging oil tankers and pipelines. To address the difficult issue of nonpoint sources, the author says that federal, state, and local partnerships are needed to monitor the release of pollutants. And finally, she notes that "since we are all in this together," we must look toward energy solutions that promote conservation, not waste, of the resource.

—The Editor

Oil in the Sea
by Nancy Rabalais

"When it rains, it pours"—or so a motorist caught in a sudden storm might think while sliding into another vehicle. It is not merely the reduced visibility and the frenetic behavior of drivers in the rain that foster such mishaps; the streets also are slicker just after the rain begins to fall. Why? Because the oil and grease that are dripped, spewed, or otherwise inadvertently deposited by motor vehicles onto roadways are among the first materials to be lifted off by the rain, thereby literally lubricating the surface. Nor do matters end with making life miserable for motorists. The oil and grease washed off roads will most likely run into storm sewers and be discharged into the nearest body of water. From there, the oily materials often are carried to the sea, where they can cause a host of environmental problems.

These events on a rainy day in the city illustrate an important but often overlooked route by which petroleum finds its way into coastal waters. Shutting down this and other routes presents a pressing challenge. True, the nation is doing a better job than ever of keeping oil out of the marine environment. But much work remains. We need to better understand the various pathways by which oil gets into the environment, how it behaves when it gets there, what effects it has on living organisms, and, perhaps most important, what steps can be taken to further reduce the amount of petroleum that enters the nation's and the world's oceans.

SOURCES AND PROBLEMS

Approximately 75 million gallons of petroleum find their way into North America's oceans each year, according to *Oil in the Sea III: Inputs, Fates, and Effects*, a report issued in 2002 by the National Research Council (NRC). About 62 percent of the total—roughly 47 million gallons per year—derives naturally from out of the ocean floor. The rest comes from human activities.

Contrary to common belief, the bulk of human-related inputs is not due to large-scale spills and accidents that occur during the transport of crude oil or petroleum products. Indeed, these types of releases account for only about 10 percent of the oil that reaches the sea as a result of human activity. The other 90 percent comes in the form of chronic low-level releases associated with the extraction and consumption of petroleum. Within this category, the biggest problem is non-point source pollution. Rivers and streams that receive runoff from a variety of land-based activities deliver roughly 16 million gallons of oil to North American coastal waters each year, more than half of the total anthropogenic load. The loads are most obvious in watersheds that drain heavily populated areas. Other sources of oil that turns up in the marine environment include jettisoned aircraft fuel, marine recreational vehicles, and operational discharges, such as cargo washings and releases from petroleum extraction.

There is at least some good news. Less oil is now entering the oceans as compared to the levels found in a previous NRC report issued in 1985. Some of this change may be attributable to differences between methodologies used in the two reports, but some decreases are due to improved regulations regarding how oil is produced and shipped. Spills from vessels in North American waters from 1990 through 1999 were down by nearly two-thirds compared to the prior decade. There also has been a dramatic decline in the amount of oil released into the environment during exploration for and production of petroleum and natural gas. Still, the recent NRC report concludes that despite such progress, the damage from oil in the marine environment is considerably more pervasive and longer-term than was previously understood.

Oil in the sea, whether from catastrophic spills or chronic releases, poses a range of environmental problems. Major spills receive considerable public attention because of the obvious attendant environmental damage, including oil-coated shorelines and dead or moribund wildlife, especially among seabirds and marine mammals. The largest oil spill in U.S. waters occurred on March 24, 1989, when the tanker *Exxon Valdez*, en route from Valdez, Alaska, to Los Angeles, California, ran aground on Bligh Reef in Prince William Sound, Alaska. Within six hours of the grounding, the ship spilled approximately 10.9 million gallons of crude oil, which would eventually affect more than 1,100 miles [1,770 km] of coastline. Large numbers of animals were killed directly, including an estimated 900 bald eagles, 250,000 seabirds, 2,800 sea otters, and 300 harbor seals.

Oil pollution also can have more subtle biological effects, caused by the toxicity of many of the compounds contained in petroleum or by the toxicity of compounds that form as the petroleum degrades over time. These effects may be of short duration and limited impact, or they may span long periods and affect entire populations or communities of organisms, depending on the timing and duration of the spill and the numbers and types of organisms exposed to the oil.

Of course, oil spills need not be large to be hazardous to marine life. Even a small spill in an ecologically sensitive area can result in damage to individual organisms or entire populations. A spill's influence also depends on the type and amount of toxins present in the petroleum product released. For instance, the fuel oil leaked when the tanker *Prestige* broke apart off the northwest coast of Spain in 2002 was initially more toxic than the crude oil spilled from the *Exxon Valdez.*

One major problem with all spills, no matter their size or type, is that the oil can remain in the environment for a long time. Several lines of evidence point to continued exposure of marine organisms to oil spilled by the *Exxon Valdez.* Substantial subsurface oil beneath coarse beaches in the spill area was found in the summer of 2001. The oil was still toxic and appeared to be chemically unchanged since its release more than a decade earlier. Some researchers predict that oil beneath mussel beds in the region affected by the spill may not decline to background levels for at least another two decades. In another instance, scientists studying salt marshes in Massachusetts that had been covered by fuel oil spilled from the barge *Florida* 30 years ago recently reported that oil is still present in sediments at depths of 6 to 28 centimeters [2.4 to 11 inches]. Moreover, the concentrations of oil found in the sediments are similar to those observed shortly after the spill. The researchers, from the Woods Hole Oceanographic Institution, predict that the compounds may remain there indefinitely, while crabs and other intertidal organisms continue to burrow through the oil-contaminated layer.

PRESCRIBED REMEDIES

Reducing the threat of oil in the oceans will require blocking the routes by which oil enters the environment. Focusing on inputs from spills and nonpoint sources, two of the major anthropogenic contributors, will show the range of actions needed.

Reducing Spills

Worldwide, large spills resulting from tanker accidents are down considerably from the totals reported by the NRC in 1985—they have decreased to 17 million gallons annually from 140 million gallons annually. This gain was achieved even as the size of the global tanker fleet increased by 900 vessels, to a total of 7,270 in 1999. Progress was made through the implementation of numerous regulations and by technological advances in vessel construction, including the increased use of double-hull tankers, the use of new construction materials, and improvements in vessel design. Spills larger than 50,000 gallons now represent less than 1 percent of total spills by number but are responsible for more than 80 percent of the total spill volume. It is important to note, however, that more than half of all tanker spills now occur in North American waters. Although the number and size of spills in these waters have been reduced considerably during the past two decades, with total volume falling to 2.5 million gallons per year, they remain the dominant domestic source of oil input to the marine environment from petroleum transportation activities, as they are globally.

Prevention, in the form of stricter regulations for tankers, has obviously not prevented all large spills, as the *Prestige* spill so dramatically illustrated. Irreparably damaged by a storm, the ship spilled nearly 3 million gallons of fuel oil, which spread over 125 miles [201 km] of coastline in one of Spain's leading areas of commercial fishing and shellfishing. But out of this calamity may come improved policies. The spill has highlighted concerns about older, single-hull ships (the *Prestige* was 26 years old) that are due to be phased out by 2015, and about what Europe should do to keep these ships safe and inspected in the meantime. Under a proposal made by the European Commission in December 2002, which the European Union (EU) is expected to adopt, single-hull oil tankers will not be allowed to carry heavy grades of oil in EU waters. Prohibited grades will include heavy fuel oil, heavy crude oil, waste oils, bitumen, and tar. Questions also have been raised about single-hull ships that are bypassing EU ports in order to avoid tough new EU-mandated inspection rules adopted in 1999 after the *Erika* oil spill polluted 250 miles [402 km] of French shoreline. The EU regulations before the passage of the latest restrictions on transport in EU waters required port authorities to check at least 25 percent of all ships coming into dock, starting with older, single-hull vessels, with priority going to ships flying flags of convenience or registered in countries with lax safety, labor, or tax rules.

The potential for a large tanker spill, however, is still significant, especially in regions without stringent safety procedures and maritime inspection practices. Furthermore, tanker traffic is expected to grow over the coming decades, as the centers of oil production continue to migrate toward the Middle East, Russia, and former Soviet states. U.S. agencies should expand their efforts to work with ship owners, domestically and internationally, through the International Maritime Organization, to enforce and build on the international regulatory standards that have contributed to the recent decline in oil spills and operational discharges.

Tankers are not the only potential source of large spills. There also is concern about aging oil pipelines and other coastal facilities. The aging of the infrastructure in fields in the central and western Gulf of Mexico and in some areas of Alaska is especially disconcerting, because these facilities often lie near sensitive coastal areas. Many pipelines in coastal Louisiana that should be buried no longer are. Numerous wellheads and other facilities within the estuaries are being abandoned, as one company takes over facilities from another. As the resources become depleted, the cost of extraction exceeds the profit to be gained from sale of the product, and owners file bankruptcy and abandon holdings. Federal agencies should work with state environmental agencies and industry to evaluate the threat posed by aging pipelines and abandoned facilities, and to take steps to minimize the potential for spills.

Reducing Nonpoint Source Inputs

Regulations developed under the Clean Water Act of 1972 have significantly reduced the number and amount of pollutants coming from the end of a pipe, and the Toxics Release Inventory tracks many of the pollutants that are released. But the more diffuse sources, such as urban runoff, atmospheric deposition, and watershed drainage, are not regulated or even monitored adequately. Unfortunately, nonpoint source runoff is difficult to measure and sparsely sampled; as a result, estimates have a high degree of uncertainty. Clearly, new federal, state, and local partnerships are needed to monitor runoff and to keep better track of how much petroleum and other pollutants industry and consumers are releasing.

Such a call for increased monitoring will undoubtedly elicit groans from managers and other people responsible for water quality, yet few people would argue that existing efforts are adequate for the overall task. Even the better-funded federal efforts are insufficient. The National Stream Quality Accounting Network, the national network operated by the U.S. Geological Survey to monitor water quality in streams, has increasingly fewer stations, particularly along the coast, as budgets are tightened and tightened again. Additional funding is necessary to invigorate this program. Coastal stations that have been shut down need to be restored, and new stations along the coast and at critical inland locations need to be added. The network also must expand monitoring to include total hydrocarbons (instead of merely "oil and grease," as is now the case), as well as a particular class of compounds called polynuclear aromatic hydrocarbons (PAHs). Growing evidence indicates that even at very low concentrations, PAHs carried in crude oil or refined products can have adverse effects on biota. This suggests that PAHs released from chronic sources may be of greater concern than previously recognized, and that in some instances the effects of petroleum spills may last longer than expected.

There also is a great need for expanded basic research. The most significant unanswered questions remain those regarding the effects on ecosystems of long-term, chronic, low-level exposures resulting from petroleum discharges and spills caused by development activities. Federal agencies, especially the Environmental Protection Agency (EPA), the U.S. Geological Survey, and the National Oceanic and Atmospheric Administration, should work with academia and industry to develop and implement a major research effort to more fully understand and evaluate the risk posed to organisms and the marine environment by the chronic release of petroleum, especially the cumulative effects of chronic releases and multiple types of hydrocarbons.

Apace with advances in monitoring and research, positive steps can be taken to implement proven methods for reducing nonpoint source discharges of oil into the environment. Remember, for example, the motorist brought low by slick pavement. The oil and grease that wash off highways during rain storms usually bypass sewage treatment plants in storm water overflow systems that pump the rain and any materials caught up in the flow directly to the closest body of water. In many urban settings, this source can be a significant contribution of petroleum to the ocean. As the population of coastal regions increases, urban runoff will become more polluted because of the expansion in the numbers of cars, asphalt-covered highways and parking lots, municipal wastewater loads, and the use and improper disposal of petroleum products. Collection and treatment of storm water overflows is necessary to control these inputs. Improved landscape management and urban management, increasing use of fuel-efficient vehicles, and public education can all contribute to lessening petroleum runoff.

The power of public education can be seen in the Chesapeake Bay region. In small but effective ways, people living within the bay's watershed are reminded daily of their consumptive uses of pollutants that enter the water. They see license plates that proclaim "Save the Bay" and storm water drain covers that say "Don't Dump. Drains to Chesapeake Bay." Education is a first step for a better-informed public that will recognize the need for less consumption, less pollution, and better conservation of resources. This knowledge should but does not always lead to legislation and funding for reducing pollutant loads, including oil reaching the sea.

Additional remedial actions should target the recreational watercraft that have grown so popular during the past two decades. Most of these craft, including jet skis and small boats with outboard motors, use two-stroke engines, which release up to 30 percent of their fuel directly into the water. Collectively, these watercraft contribute almost 1 million gallons of petroleum each year into North American waters. The bulk of their input is in the form of gasoline, which is thought to evaporate rapidly from the water surface. However, little is known about the actual fate of the discharge, or about its biological effects while in its volatile phase, which is highly toxic. In 1990, heightened awareness of this problem led the EPA to begin regulating the "nonroad engine" population, under the authority of the Clean Air Act. Questions remain regarding the amount of petroleum residing in the water column or along the surface for biologically significant lengths of time. The EPA should continue its phase-out efforts directed at two-stroke engines, and it should expand research, in conjunction with other relevant federal agencies, on the fate and effects of discharges from these older, inefficient motors.

To achieve maximal effectiveness, efforts to understand and minimize oil pollution should pay heed to worldwide needs. The United States and other developed countries have invested much in technologies to reduce the spillage of oil into the marine environment, as well as in the science that has increased knowledge of the effects of spilled oil, either acute or chronic. This knowledge should be transferred to people in other countries that are developing their petroleum reserves. It is imperative that the petroleum industry not simply comply with regulations in developing countries where they operate, but that they transfer the knowledge derived from extensive studies in U.S. waters to areas where their operations are expanding.

SHARED RESPONSIBILITIES

It is tempting to blame the oil and shipping industries alone for spills such as those from the *Exxon Valdez* and the *Prestige*, but everyone who benefits from oil bears responsibility for the fraction that enters the sea. If companies have failed to build and buy double-hull tankers, it is in part because consumers do not wish to pay the increased fuel prices that would be needed to offset the extra cost. The push for improved methods of extracting, producing, and transporting oil must come from the general public, and this link reinforces the need for education.

The price of oil and natural gas is a major force in the world economy. As recently as the late 1990s, the average price of a barrel of crude oil was less than the cost of a takeout dinner. Yet a fluctuation of 20 or 30 percent in the price can influence automotive sales, travel decisions, interest rates, stock market trends, and the gross national products of industrialized nations. Perceived or real decreases in the availability of oil led to long lines for gasoline in the early 1970s and to the development and sale of fuel-efficient vehicles. Many observers argue that the low oil prices in recent years have helped put a glut of gas-guzzling vehicles on the highway. As the prices of a barrel of oil and a gallon of gasoline continue to rise in the face of social unrest in South America and the unrelenting hostilities in the Middle East, the value of this limited commodity may become more apparent.

The United States needs an energy policy that treats petroleum as a limited and treasured commodity that encourages conservation rather than waste, and that supports alternative energy development and use. The nation also should tighten controls on air and water pollution and should adequately fund environmental monitoring of water resources. Without these policy changes, U.S. citizens cannot expect wise and environmentally sound use of the nation's or the world's resources. And if U.S. citizens cannot reduce their overly consumptive use of petroleum to help in curbing its introduction to the diffuse but voluminous nonpoint source pollutant load, how can they expect citizens of developing countries to strive for less polluting and consumptive actions in the face of an improving economy derived from sales of petroleum?

"We're all in this together," as the saying goes, including the motorist who crashed on that oil-slicked road. There is just one global economy, one global ecosystem, and one global source of nonrenewable petroleum reserves.

Why Are Plastics in the Ocean a Growing Hazard?

Many of the things we buy or use are made from or packaged in plastic. One environmental problem this fact presents is that bacteria cannot degrade or decompose plastics. Most plastics are photodegradable, meaning that they break down only in the presence of light. The author of the following article, Charles Moore, discovered a startling patch of plastic garbage in the middle of the Pacific Ocean—startling because of its location in a place where few ships pass and because it was nearly the size of the state of Texas!

Large pieces and tiny pellets of plastic swirl in the ocean currents. Plastic is ingested by organisms, and it kills them. For example, sea turtles often eat plastic bags, thinking they are jellyfish. Or animals may strangle in plastic materials. But there is another problem that we cannot see. It has been discovered that these plastic polymers absorb and concentrate toxic chemicals. So as plastics degrade further and travel through ocean waters, they spread toxins. These toxins are eaten by plankton, which is in turn eaten by small fish, which are eaten by larger fish, mammals, and birds. In this way, the toxins become concentrated in larger animals. The problem is no longer just having unsightly plastic garbage floating in the middle of the ocean. The problem is the harm being done to fish, birds, mammals, and other creatures of the sea.

The research conducted by Charles Moore and the scientific groups with which he works is extremely important because it provides data that can be used by policymakers who are concerned with protecting ocean resources. The author is captain of the *Alguita*, an independent oceanographic research vessel. In 1994, Moore founded the Algalita Marine Research Foundation (*www.algalita.org*), which is dedicated to research, education, and restoration of the marine environment.

—The Editor

Trashed: Across the Pacific Oceans, Plastics, Plastics Everywhere
by Charles Moore

It was on our way home, after finishing the Los Angeles-to-Hawaii sail race known as the Transpac, that my crew and I first caught sight of the trash, floating in one of the most remote regions of all the oceans. I had entered my cutter-rigged research vessel, *Alguita*, an aluminum-hulled catamaran, in the race to test a new mast. Although *Alguita* was built for research trawling, she was also a smart sailor, and she fit into the "cruising class" of boats that regularly enter the race. We did well, hitting a top speed of twenty knots under sail and winning a trophy for finishing in third place.

Throughout the race our strategy, like that of every other boat in the race, had been mainly to avoid the North Pacific subtropical gyre—the great high-pressure system in the central Pacific Ocean that, most of the time, is centered just north of the racecourse and halfway between Hawaii and the mainland. But after our success with the race we were feeling mellow and unhurried, and our vessel was equipped with auxiliary twin diesels and carried an extra supply of fuel. So on the way back to our home port in Long Beach, California, we decided to take a shortcut through the gyre, which few seafarers ever cross. Fishermen shun it because its waters lack the nutrients to support an abundant catch. Sailors dodge it because it lacks the wind to propel their sailboats.

I often struggle to find words that will communicate the vastness of the Pacific Ocean to people who have never been to sea. Day after day, *Alguita* was the only vehicle on a highway without landmarks, stretching from horizon to horizon. Yet as I gazed from the deck at the surface of what ought to have been a pristine ocean, I was confronted, as far as the eye could see, with the sight of plastic.

It seemed unbelievable, but I never found a clear spot. In the week it took to cross the subtropical high, no matter what time of day I looked, plastic debris was floating everywhere: bottles, bottle caps, wrappers, fragments. Months later, after I discussed what I had seen with the oceanographer Curtis Ebbesmeyer, perhaps the world's leading expert on flotsam, he began referring to the area as the "eastern garbage patch." But "patch" doesn't begin to convey the reality. Ebbesmeyer has estimated that the area, nearly covered with floating plastic debris, is roughly the size of Texas.

My interest in marine debris did not begin with my crossing of the North Pacific subtropical gyre. Voyaging in the Pacific has been part of my life since earliest childhood. In fifty-odd years as a deckhand, stock tender, able seaman, and now captain, I became increasingly alarmed by the growth in plastic debris I was seeing. But the floating plastics in the gyre galvanized my interest.

I did a quick calculation, estimating the debris at half a pound for every hundred square meters of sea surface. Multiplied by the circular area defined by our roughly thousand-mile course through the gyre, the weight of the debris was about 3 million tons, comparable to a year's deposition at Puente Hills, Los Angeles's largest landfill. I resolved to return someday to test my alarming estimate.

Historically, the kind of drastic accumulation I encountered is a brand-new kind of despoilment. Trash has always been tossed into the seas, but it has been broken down in a fairly short time into carbon dioxide and water by marine microorganisms. Now, however, in the quest for lightweight but durable means of storing goods, we have created a class of products—plastics— that defeat even the most creative and voracious bacteria.

Unlike many discarded materials, most plastics in common use do not biodegrade. Instead they "photodegrade," a process whereby sunlight breaks them into progressively smaller pieces, all of which are still plastic polymers. In fact, the degradation eventually yields individual molecules of plastic, but these are still too tough for most anything—even such indiscriminate consumers as bacteria—to digest. And for the past fifty years or so, plastics that have made their way into the Pacific Ocean have been fragmenting and accumulating as a kind of swirling sewer in the North Pacific subtropical gyre.

It surprised me that the debris problem in the gyre had not already been looked at more closely by the scientific community. In fact, only recently—starting in the early 1990s—has the scientific community begun to focus attention on the trash in the gyre. One of the first investigators to study the problem was W. James Ingraham Jr., an oceanographer at the National Oceanic and Atmospheric Administration (NOAA) in Seattle. Ingraham's Ocean Surface Current Simulator (OSCURS) predicts that objects reaching this area might revolve around in it for sixteen years or more.

A year after my sobering voyage, I asked Steven B. Weisberg, director of the Southern California Coastal Water Research Project and an expert in marine environmental monitoring, to help me make a more rigorous estimate of the extent of the debris in the subtropical gyre. Weisberg's group had already published an article on the debris they had collected in fish trawls of the Southern California Bight, a region along the Pacific coast extending a hundred miles both north and south of Los Angeles. As I discussed the design plan for our survey with Weisberg's statisticians, Molly K. Leecaster and Shelly L. Moore, it became apparent that we were facing a new problem. In the coastal ocean, bodies of water are naturally defined, in part, by the coasts they lie against. In the open ocean, however, bodies of water are bounded by atmospheric pressure systems and the currents those systems create. In other words, air, not land, defines the body of water. Because air pressure systems move, the body of water we wanted to survey would be moving as well. A random sample of a moving area such as the gyre would have to be done quite differently from the way Weisberg's group had conducted their survey along the Pacific coast.

The gyre we planned to survey is one of the largest ocean realms on Earth, and one of five major subtropical gyres on the planet. Each subtropical gyre is created by mountainous flows of air moving from the tropics toward the polar regions. The air in the North Pacific subtropical gyre is heated at the equator and rises high into the atmosphere because of its buoyancy in cooler, surrounding air masses. The rotation of the Earth on its axis moves the heated air mass westward as it rises, then eastward once it cools and descends at around 30 degrees north latitude, creating a huge, clockwise-rotating mass of air.

The rotating air mass creates a high-pressure system throughout the region. Those high pressures depress the ocean surface, and the rotating air mass also drives a slow but oceanic-scale surface current that moves with the air in a clockwise spiral. Winds near the center of the high are light or even calm, and so they do not mix the floating debris into the water column. This huge region, what I call a "gentle maelstrom," has become an accumulator of debris from innumerable sources along the North Pacific rim, as well as from ships at sea.

The subtropical gyres are also oceanic deserts—in fact, many of the world's land-based deserts lie at nearly the same latitudes as the oceanic gyres. Like their terrestrial counter-parts, the oceanic deserts are low in biomass. On land the low biomass is caused by the lack of moisture; in oceanic deserts the low biomass is a consequence of great ocean depths.

In coastal areas and shallow seas, winds and waves con-stantly stir up and recycle nutrients, increasing the biomass of the food web. In the deep oceans, though, such forces have no effect; the bottom sequesters the nutrient-rich residue of millions of years of near-surface photosynthetic production, as well as the decomposed fragments of life in the sea, trapping them miles below the surface. Hence the major source of food for the web of life in deep ocean areas is photosynthesis.

But even in the clear waters that prevail in the subtropical gyres, photosynthesis is confined to the top of the water column. Sunlight attenuates rapidly with depth, and by the time it has gone only about 5 percent of the way to the bottom, the light is too weak to fuel marine plants. The net effect is a vast area poor in resources, an effect that makes itself felt throughout the food web. Top predators such as tuna and other commercially viable fish don't hang out in the gyres because the density of prey is so low. The human predator stays away too: the resources that have drawn entrepreneurs and scientists alike to various regions of the ocean are not present in the subtropical gyres.

What does exist in the gyres is a great variety of filter-feeding organisms that prey on the ever-renewed crop of tiny plants, or phytoplankton. Each day the phytoplankton grow in the sunlit part of the water, and each night they are consumed by the filter feeders, a fantastic array of alien-looking animals called zooplankton. The zooplankton include chordate jellyfishes known as "salps," which are among the fastest-growing multicellular organisms on the planet. By fashioning their bodies into pulsating tubes, the salps are able, each day, to filter half the water column they inhabit, drawing out the phytoplankton and smaller zooplankton for food. But salps are gelatinous creatures with a low biomass, and so there is no market for them, either. Hence the realm they dominate, one of the largest uniform habitats on the planet, remains unexploited and largely unexplored.

Leecaster, Moore, and I came up with a plan to make a series of trawls with a surface plankton net, along paths within a circle with a 564-mile [908-km] radius. The area of the circle would then be almost exactly 1 million square miles [2.6 million km²]. Trawling would start when we estimated we were under the central pressure cell of the high-pressure system that creates the gyre. We would regard the starting point as the easternmost point along the circumference of the circle. Then we would proceed due west to the center of the circle, turn south, and sail back to the southernmost point on the circumference, alternating between trawling and cruising. We intended to obtain transect samples with random lengths and random spacing between trawls. To be conservative about our sampling technique, we decided that any debris we collected would count only as a sample of the debris within the area of the transected circle.

In August 1998 I set out with a four-member volunteer crew from Point Conception, California heading northwest toward the subtropical gyre. Onboard *Alguita* was a manta trawl, an apparatus resembling a manta ray with wings and a broad mouth, which skimmed the ocean surface trailing a net with a fine mesh. Eight days out of port, the wind dropped below ten knots and we decided to practice our manta trawling technique, taking a sample at the edge of the subtropical gyre, about 800 miles [1,288 km] offshore. We pulled in the manta after trawling three and a half miles [5.6 km].

What we saw amazed us. We were looking at a rich broth of minute sea creatures mixed with hundreds of colored plastic fragments—a plastic-plankton soup. The easy pickings energized all of us, and soon we began sampling in earnest. Because plankton move up and down in the water column each day, we needed to trawl nonstop, day and night, to get representative samples. When we encountered the light winds typical of the subtropical gyre, we deployed the manta outside the port wake, along with two other kinds of nets. Each net caught plenty of debris, but far and away the most productive trawl was the manta.

There was plenty of larger debris in our path as well, which the crew members retrieved with an inflatable dingy. In the end, we took about a ton of this debris on board. The items included

- a drum of hazardous chemicals;

- an inflated volleyball, half covered in gooseneck barnacles;

- a plastic coat hanger with a swivel hook;

- a cathode-ray tube for a nineteen-inch TV;

- an inflated truck tire mounted on a steel rim;

- numerous plastic, and some glass, fishing floats;

- a gallon bleach bottle that was so brittle it crumbled in our hands; and

- a menacing medusa of tangled net lines and hawsers that we hung from the A-frame of our catamaran and named Polly P, for the polypropylene lines that made up its bulk.

In 2001, in the *Marine Pollution Bulletin*, we published the results of our survey and the analysis we had made of the debris, reporting, among other things, that there are six pounds [2.7 kg] of plastic floating in the North Pacific subtropical gyre for every pound of naturally occurring zooplankton. Our readers were as shocked as we were when we saw the yield of our first trawl. Since then we have returned to the area twice to continue documenting the phenomenon. During the latest trip, in the summer of 2002, our photographers captured underwater images of jellyfish hopelessly entangled in frayed lines, and transparent filter feeding organisms with colored plastic fragments in their bellies.

Entanglement and indigestion, however, are not the worst problems caused by the ubiquitous plastic pollution. Hideshige Takada, an environmental geochemist at Tokyo University, and his colleagues have discovered that floating plastic fragments accumulate hydrophobic—that is, non-water-soluble—toxic chemicals. Plastic polymers, it turns out, are sponges for DDT, PCBs, and other oily pollutants. The Japanese investigators found that plastic resin pellets concentrate such poisons to levels as high as a million times their concentrations in the water as free-floating substances.

The potential scope of the problem is staggering. Every year some 5.5 quadrillion (5.5 x 10^15) plastic pellets—about 250 billion pounds of them—are produced worldwide for use in the manufacture of plastic products. When those pellets or products degrade, break into fragments, and disperse, the pieces may also become concentrators and transporters of toxic chemicals in the marine environment. Thus an astronomical number of vectors for some of the most toxic pollutants known are being released into an ecosystem dominated by the most efficient natural vacuum cleaners nature ever invented: the jellies and salps living in the ocean. After those organisms ingest the toxins, they are eaten in turn by fish, and so the poisons pass into the food web that leads, in some cases, to human beings. Farmers can grow pesticide-free organic produce, but can nature still produce a pollutant-free organic fish? After what I have seen firsthand in the Pacific, I have my doubts.

Many people have seen photographs of seals trapped in nets or choked by plastic six-pack rings, or sea turtles feeding on plastic shopping bags, but the poster child for the consumption of pelagic plastic debris has to be the Laysan albatross. The plastic gadgets one typically finds in the stomach of the bird—whose range encompasses the remote, virtually uninhabited region around the northwest Hawaiian Islands—could stock the checkout counter at a convenience store. My analysis of the stomach contents of birds from two colonies of Laysan albatrosses that nest and feed in divergent areas of the North Pacific show differences in the types of plastic they eat. I believe those differences reveal something about the way plastic is transported and breaks down in the ocean.

On Midway Island in the Hawaiian chain, a bolus, or mass of chewed food, coughed up by one bird included many identifiable objects. By contrast, a bird on Guadalupe Island, which lies 150 miles [241 km] off the coast of Baja California, produced a bolus containing only plastic fragments. The principal natural prey of both bird colonies is squid, but as the ecologist Carl Safina notes in his book *Eye of the Albatross*, the birds' foraging style can be described as "better full than fussy." Robert W. Henry III, a biologist at the University of California, Santa Cruz, and his colleagues have tracked both the Hawaiian and the Guadalupe populations of birds and found that the foraging areas of each colony in the Pacific are generally nonoverlapping and wide apart.

One difference between the two areas is apparently the way debris flows into them. In Ingraham's OSCURS model, debris from the coast of Japan reaches the foraging area of the Hawaiian birds within a year. Debris from the West Coast of the United States, however, sticks close to the coast until it bypasses the foraging area of the Guadalupe birds, then heads westward to Asia, not to return for six years or more. The lengthy passage seems to give the plastic debris time to break into fragments.

The subtropical gyres of the world are part of the deep ocean realm, whose ability to absorb, hide, and recycle refuse has long been seen as limitless. That ecologically sound image, however, was born in an era devoid of petroleum-based plastic polymers. Yet the many benefits of modern society's productivity have made nearly all of us hopelessly, and to a large degree rationally, addicted to plastic. Many, if not most, of the products we use daily contain or are contained by plastic. Plastic wraps, packaging, and even clothing defeat air and moisture and so defeat bacterial and oxidative decay. Plastic is ubiquitous precisely because it is so good at preventing nature from robbing us of our hard-earned goods through incessant decay.

But the plastic polymers commonly used in consumer products, even as single molecules of plastic, are indigestible by any known organism. Even those single molecules must be further degraded by sunlight or slow oxidative breakdown before their constituents can be recycled into the building blocks of life. There is no data on how long such recycling takes in the ocean—some ecologists have made estimates of 500 years or more. Even more ominously, no one knows the ultimate consequences of the worldwide dispersion of plastic fragments that can concentrate the toxic chemicals already present in the world's oceans.

Ironically, the debris is re-entering the oceans whence it came; the ancient plankton that once floated on Earth's primordial sea gave rise to the petroleum now being transformed into plastic polymers. That exhumed life, our "civilized plankton," is, in effect, competing with its natural counterparts, as well as with those life-forms that directly or indirectly feed on them.

And the scale of the phenomenon is astounding. I now believe plastic debris to be the most common surface feature of the world's oceans. Because 40 percent of the oceans are classified as subtropical gyres, a fourth of the planet's surface area has become an accumulator of floating plastic debris. What can be done with this new class of products made specifically to defeat natural recycling? How can the dictum "In ecosystems, everything is used" be made to work with plastic?

BIBLIOGRAPHY

American Rivers and Sierra Club. *Where Rivers Are Born: The Scientific Imperative for Defending Small Streams and Wetlands.* 2003. Available online at *http://www.amrivers.org/doc_repository/ WhereRviersAreBorn1.pdf.*

Bennett, Elena, and Steve Carpenter. "P Soup: The Global Phosphorous Cycle." *World Watch Institute.* March 2002. Available online at *http://www.worldwatch.org/pubs/mag/2002/152/.*

Dudley, Nigel, and Sue Stolton. *Running Pure: The Importance of Forest Protected Areas to Drinking Water.* World Bank/World Wildlife Fund, 2003.

Ebersole, Rene. "Is Your Drinking Water Safe?" *National Wildlife Federation.* July 2004. Available online at *http://www.nwf.org/nationalwildlife/ article.cfm?articleId=937&issueId=68.*

Moore, Charles. "Trashed: Across the Pacific Oceans, Plastics, Plastics Everywhere." *Natural History.* November 2003. Available online at *http://www.naturalhistorymag.com/.*

Natural Resources Defense Council. *Testing the Waters 2004: A Guide to Water Quality at Vacation Beaches.* Available online at *http://www.nrdc.org/.*

Orlins, Joseph, and Anner Wehrly. "The Quest for Clean Water." *World & I.* May 2003.

Pew Oceans Commission. *America's Living Oceans: Charting a Course for Sea Change.* May 2003. Available online at *http://www.pewoceans.org/ oceans/pew_oceans_report_c5.asp.*

Rabalais, Nancy. "Oil in the Sea." *Issues in Science and Technology.* Fall 2003. Available online at *http://www.issues.org/issues/20.1/rabalais.html.*

Sampat, Payla. "Groundwater Shock." *World Watch Institute Magazine.* January–February 2000. Available online at *http://worldwatch.org/ pubs/mag/2000/131/.*

Statchell, Michael. "Troubled Waters." *National Wildlife Federation.* February–March 2003. Available online at *http://www.nwf.org/ nationalwildlife/dspPlainText.cfm?articleId=731.*

U.S. Environmental Protection Agency. *It's Your Drinking Water: Get to Know It and Protect It.* Available online at *http://www.epa.gov/safewater/consumer/itsyours.pdf.*

U.S. Geological Survey Report Survey #1265. *Water Quality in the Nation's Streams and Aquifers.* 2004. Available online at *http://water.usgs.gov/ pubs/circ/2004/1265/pdf/circular1265.pdf.*

FURTHER READING

Dowie, Mark. *Losing Ground: American Environmentalism at the Close of the Twentieth Century.* Cambridge, MA: MIT Press, 1995.

Leopold, Aldo. *A Sand County Almanac.* New York: Oxford University Press, 1949.

Turco, Richard P. *Earth Under Siege: From Air Pollution to Global Change.* New York: Oxford University Press, 2002.

WEBSITES

American Rivers and Sierra Club
http://www.amrivers.org/

National Wildlife Federation
http://www.nwf.org/

Natural Resources Defense Council
http://www.nrdc.org/

Pew Oceans Commission
http://www.pewoceans.org/

U.S. Geological Survey
http://water.usgs.gov/

World Watch Institute
http://worldwatch.org

INDEX

INDEX

INDEX

INDEX

INDEX

ABOUT THE CONTRIBUTORS

YAEL CALHOUN is a graduate of Brown University and received her M.A. in Education and her M.S. in Natural Resources Science. Years of work as an environmental planner have provided her with much experience in environmental issues at the local, state, and federal levels. Currently she is writing books, teaching college, and living with her family at the foot of the Rocky Mountains in Utah.

Since 2001, DAVID SEIDEMAN has served as editor-in-chief of *Audubon* magazine, where he has worked as an editor since 1996. He has also covered the environment on staff as a reporter and editor for *Time, The New Republic,* and *National Wildlife.* He is the author of a prize-winning book, *Showdown at Opal Creek,* about the spotted owl conflict in the Northwest.